Hemal Lining Defaillance in Zymotic Ensemic and Glycometabolic Nosopathy With and Without Sequela

Compita Signora

Hemal Lining Defaillance in Zymotic Ensemic and Glycometabolic Nosopathy With and Without Sequela

First Edition August 2025

Written by Compita Signora

Table of Contents

S.No	Content	Page No
	Chapter 1. Introduction	01
	Chapter 2. Review of literature	05
	Part I - Diabetes Mellitus	06
2.1	Types of diabetes mellitus	07
2.1.1	Classification of Type 2 Diabetes	07
2.1.2	Etiologic Classification of Diabetes Mellitus	09
2.1.3	Other types of diabetes	10
2.1.4	Gestational diabetes mellitus (GDM)	11
2.2	Epidemiology of diabetes mellitus	11
2.2.1	Global prevalence of diabetes mellitus	11
2.2.2	Evolution of the diabetes epidemic in India	14
2.3	Historical aspects of diabetes	18
2.3.1	History of diabetes in timeline	19
2.4	Pathophysiology of type 2 diabetes	23
2.4.1	Natural history of type 2 diabetes and pathophysiology	23
2.5	Metabolic changes during the development of type 2 diabetes mellitus	25

Table of Content

S. NO	Content	Page no
2.5.1	Abnormal muscle and fat metabolism	26
2.5.2	Impaired insulin secretion	27
2.5.3	Increased hepatic glucose and lipid production	28
2.6	Glucose homeostasis	28
2.6.1	Definition of glucose homeostasis	28
2.6.2	Mechanisms of glucose homeostasis	29
2.7	Insulin	30
2.7.1	Discovery of Insulin	30
2.7.2	Structure and chemical properties of Insulin	30
2.7.3	Synthesis and release of Insulin	31
2.7.4	Factors influencing β cell mass	33
2.8	Physiology of insulin secretion and response to various stimuli	34
2.8.1	Response to glucose	35
2.8.2	Response to arginine	35
2.8.3	Effects of lipids	35
2.8.4	Insulin receptors and insulin binding	36
2.8.5	Glucose transporter proteins	37
2.8.6	Actions of insulin on different metabolisms	38
	A. Carbohydrate metabolism	38
	B. Lipid metabolism	39
	C. Protein synthesis	39
2.9	Diagnostic criteria for diabetes mellitus	40
2.9.1	WHO (2002) criteria for the diagnosis of diabetes mellitus	40

Table of Content

S. NO	Content	Page no
2.9.2	Diagnostic indices of diabetes mellitus	40
	A. Glucose determination	40
	B. Glycated proteins	44
	C. Estimation of lipid profile	47
	D. Glucose tolerance test	50
	E. Other indices	52
	F. Laboratory tests for glycemic control	54
	Part II- Metabolic Syndrome	57
2.10	What is Pre-diabetes?	58
2.10.1	Metabolic syndrome	59
2.10.2	Pathophysiology of Metabolic Syndrome i) Insulin resistance: a conceptual prologue ii) Obesity as a "driving force" in the prevalence of insulin Resistance iii) Insulin resistance in adipose tissue iv) Insulin resistance in the liver v) Insulin resistance in muscle vi) Hypertension and insulin resistance vii) Should pre-diabetes be called a disease?	60
2.10.3	Diagnostic criteria for metabolic syndrome	66
2.10.4	Dignostic methods for assessing insulin resistance and β-cell function **i)** Hyperinsulinemic euglycemic glucose clamp **ii)** Homeostasis Model of Assessment **iii)** QUICKI **iv)** Surrogate Markers	67
	Part III- Diabetic complications and endothelial dysfunction	72
2.11	Diabetic complications and their Pathogenesis	73
2.11.1	Diabetic retinopathy	74
2.11.2	Diabetic neuropathy	74

Table of Content

S. NO	Content	Page no
2.11.3	Diabetic nephropathy	74
2.11.4	Cardiovascular morbidity and mortality	75
2.11.5	Vascular complications in diabetes mellitus	75
2.12	Endothelial function and dysfunction	76
2.12.1	Nitric oxide	79
2.12.2	Nitric oxide synthase enzyme	80
2.12.3	Isoenzymes of nitric oxide synthase	81
2.12.4	Chromosomal location of eNOS gene	83
2.12.5	Regulation of eNOS activity and gene expression	84
2.12.6	Association between the G894T variant of the eNOS gene and cardiovascular phenotypes	86
	Chapter 3.Hypothesis	98
	Chapter 4.Aims and objectives	100
	Chapter 5. Material and Methods	103
5.1.	Study design	104
5.1.1	Ethical clearance	104
5.1.2	Place of study	104
5.1.3	Study population	104
5.1.4	Enrollment of patients	104
5.1.5	Sampling technique	105
5.1.6.	Study procedures	105
5.1.7	Sample Size determination	105
5.2	Division of subjects in to four Groups	105
5.2.1	Data collection	108
5.3	Laboratory analysis	108
5.3.1	Anthropometric parameters A. Weight and height measurements B. Waist to hip ratio (WHR) C. Blood pressure measurements	108

Table of Content

S. NO	Content	Page no
5.3.2	**Glycemic Parameters** A. Estimation of plasma glucose by glucose oxidase and peroxidase method B. Estimation of fasting plasma insulin C. Estimation of HbA1c	109
5.3.3	**Lipid parameter** A. Estimation of total cholesterol by Cholesterol Esterase and cholesterol oxidase (CHO & POD) enzymatic method. B. Estimation triglyceride by glycerol phosphate oxidase & peroxidase (GPO-POD) in serum by enzymatic method. C. Estimation of serum HDL by cholesterol esterase and choesterol oxidase immunoinhibition method (CHE and CHO immunoinhibition method) D. Calculation of LDL cholesterol and VLDL cholesterol E. Estimation of apolipoprotein B in serum by Immunoturbidometry method	116
5.3.4	**Endothelial parameter** A. Estimation of microalbumin by immuno turbidimetric method B. Estimation of nitric oxide by modified Griess method	121
5.3.5	**Genetic Analysis** A. Extraction of human genomic DNA from blood containing EDTA. B. Qualitative, Quantitative & Integrity analysis of DNA C. Polymerase Chain Reaction i) PCR Primers designing ii) Tm Calculator results of primers for PCR iii) Procedure for amplification of DNA segment by PCR. iv) PCR conditions D. Restriction fragment length polymorphism i)Determination of restriction enzyme for polymorphism by RFLP	124
	Chapter 6. Result	144
6.1	Anthropometric measurements	146
6.2	Glycemic parameters	156
6.3	Lipid parameters	162

Table of Content

S. NO	Content	Page no
6.4	Endothelial parameters	173
6.5	Genetic parameters	184
7.1	**Chapter 7.Discussion**	191
7.1	Anthropometric parameters	194
7.1.1	Comparison of BMI and WHR among different study groups and their correlation with HOMA IR i) Group I and Group II ii) Group I and Group III iii) Group I and Group IV iv) Group II with Group III and Group IV v) Group III and Group IV	195
7.2	Blood pressure	199
7.2.1	Comparison of Blood pressure among different study groups i) Group I and Group II ii) Group I and Group III iii) Group I and Group IV iv) Group II and Group III and Group IV v) Group III and Group IV	199
7.3	Glycemic parameters	202
7.3.1	Comparison of FPG, PPPG, FPI, HbA1c and HOMA IR among the study Groups i) Group I and Group II ii) Group I & Group III iii) Group I & Group IV iv) Group II and Groups III and Group IV v) Group III and Group IV	202
7.4	Lipid parameters	209
7.4.1	Comparison of Lipid parameters among study groups and correlation with HOMA IR i) Group I and Groups II ii) Group I and Groups III iii) Group I and Group IV iv) Group II with Group III and Group IV v) Group III with Group IV	209

Table of Content

S. NO	Content	Page no
7.5	Endothelial parameters	216
7.5.1	Comparison of Endothelial markers among all the study Groups i) Group I and Group II ii) Group I and Group III iii) Group I and Group IV iv) Group II with Group III and Groups IV v) Group III & Group IV	216
7.6	Genetic analysis	215
7.6.1	Flowchart of genetic analysis	217
7.6.2	Genotype Frequency i) Genotype frequency in Control Group ii) Genotype frequency in Study Groups iii) Allelic frequency in Control Group iv) Allelic frequency in Study Group	228
7.6.3	Risk of occurrence of diseases	231
7.6.4	Genotype:Phenotype Association	231
	i) Association with serum Nitric oxide levels in different study groups	232
	➢ Within Group analysis	233
	➢ Intergroup analysis for serum nitric oxide - Group I with Group II - Group II with Group III - Group III with Group IV	234
	ii) Association of microalbuminuria	234
	➢ Within group analysis	235
	➢ Inter group analysis for microalbuminuria	235
	iii) Association of Genotype with HOMA IR	235
	➢ Within Group analysis	235
	➢ Inter Group analysis - Group I with Group II - Group II with Group III -Group III with Group IV	236
7.6.5	Conclusion of genetic analysis	237
8	Chapter 8. Summary and Conclusion	238
9.	Bibliography	242

List of Tables

Table no	Name of Table	Page No.
2.1	Top 10 countries for estimated numbers of adults with diabetes, 2010 and 2030	14
2.2	Different methods used for estimation of HbA1c	45
2.3	Reference intervals of HbA1c	46
2.4	Criteria for the metabolic syndrome definitions	66
2.5	Isoforms of nitric oxide synthase	82
2.6	Molecular biology of human isoenzymes of nitric oxide synthase	84
6.1	Anthropometric Parameters in different study Groups	147
6.1a	Comparison of Anthropometric parameters of Group I with Group II	148
6.1b	Comparison of Anthropometric parameters of Group I with Group III	148
6.1c	Comparison of Anthropometric parameters of Group I with Group IV	149
6.1d	Comparison of Anthropometric parameters of Group II with Group III	149
6.1e	Comparison of Anthropometric parameters of Group II with Group IV	150
6.1f	Comparison of Anthropometric parameters of Group III and Group IV	150
6.1g	Linear regression model between HOMA IR and WHR and BMI in Group II	154
6.1h	Linear regression model between HOMA IR and WHR and BMI in Group III	154
6.1i	Linear regression model between HOMA IR and WHR and BMI in Group IV	155
6.1j	Correlation coefficients of HOMA IR and WHR in all Groups	155
6.1k	Correlation coefficients of HOMA IR and BMI in all Groups	155
6.2	Glycemic Parameters in different study Groups	157
6.2a	Comparison of Glycemic parameters of Group I with Group II	159
6.2b	Comparison of Glycemic parameters of Group I with Group III	159
6.2c	Comparison of Glycemic parameters of Group I with Group IV	160
6.2d	Comparison of Glycemic parameters of Group II with Group III	160
6.2e	Comparison of Glycemic parameters of Group II with Group IV	161
6.2f	Comparison of Glycemic parameters of Group III and Group IV	161
6.3	Lipid Parameters in different study Groups	163
6.3a	Comparison of Lipid parameters in Group I with Group II	165
6.3b	Comparison of Lipid parameters in Group I with Group III	165
6.3c	Comparison of Lipid parameter in Group I with Group IV	166
6.3d	Comparison of Lipid parameter in Group II with Group III	166
6.3e	Comparison of Lipid parameter in Group II with Group IV	167
6.3f	Comparison of Lipid parameter in Group III and Group IV	167

List of Tables

Table no	Name of Table	Page No.
6.3g	Linear regression model between HOMA IR and TG/HDL-C ratio and Apo B in Group II	171
6.3h	Linear regression model between HOMA IR and TG/HDL-C ratio and Apo B in Group III	171
6.3i	Linear rRegression model between HOMA IR and TG/HDL-C ratio and Apo B in Group IV	172
6.3j	Correlation coefficient of HOMA IR and TG/HDL-C in all Groups	172
6.3k	Correlation coefficient of HOMA IR and Apo B in all Groups	172
6.4	Endothelial parameters in different study Groups	174
6.4a	Comparison of Endothelial parameters in Group I & Group II	176
6.4b	Comparison of Endothelial parameters in Group I & Group III	176
6.4c	Comparison of Endothelial parameters in Group I & Group IV	177
6.4d	Comparison of Endothelial parameters in Group II & Group III	177
6.4e	Comparison of Endothelial parameters in Group II & Group IV	178
6.4f	Comparison of Endothelial parameters in Group III & Group IV	178
6.4g	Linear regression model between HOMA IR and serum nitric oxide and urinary micoalbumin Group II	182
6.4h	Linear regression model between HOMA IR and serum nitric oxide and urinary micoalbumin Group III	182
6.4i	Linear regression model between HOMA IR and serum nitric oxide and urinary microalbumin Group IV	183
6.4j	Correlation coefficient of HOMA IR and serum nitric oxide in all Groups	183
6.4k	Correlation coefficient of HOMA IR and urinary microalbumin in all Groups	183
6.5a	Genotype frequency of SNP at G894T	185
6.5b	Allelic frequency of SNP at G894T	185
6.5c	Association of eNOS gene polymorphism with disease	187
6.5d	Genotype Association study	187
6.5e	Within Group pair wise Comparison	188
6.5f	Between Group pairwise comparison	188

List of Figure

Figure No.	Name of Figure	Page No.
2.1	Spectrum of glucose homeostasis and diabetes mellitus (DM)	08
2.2	Worldwide prevalence of diabetes mellitus	12
2.3	Prevalence of type 2 diabetes in different parts of India.	16
2.4	Prevalence of diagnosed and undiagnosed Diabetes and hypertension in India	17
2.5	Prevalence of diabetes and pre diabetes in urban and rural India	17
2.6	Age- and sex-specific prevalence of diabetes and impaired glucose regulation in 11 Asian cohorts	17
2.7	Natural history of type 2 diabetes mellitus	24
2.8	Metabolic changes during the development of type 2 diabetes mellitus (DM)	25
2.9	Schematic presentation of insulin secretory pathways.	34
2.10	Schematic presentation of insulin signaling pathways	37
2.11	The relationship between metabolic syndrome, insulin resistance, hyperinsulinemia and hyperglycemia (overt type 2 diabetes).	65
2.12	Obesity/insulin resistance and endothelial dysfunction closely coexist throughout the natural history of type 2 diabetes	77
2.13	Mechanisms through which obesity, insulin resistance, and endothelial dysfunction are closely associated	78
2.14	Nitric oxide synthase: models and mechanisms	80
2.15	Domain structure of nitric oxide synthases (NOS)	81
2.16	Location of eNOS gene on chromosome 7	83
5.1	Plasma Insulin kit by Immunoshop	113
5.2a	ELISA assay for Insulin after addition of chromogenic solution	113
5.2b	ELISA assay for Insulin after addition of stop solution	114
5.3	DNA Gel electrophoresis (Integrity check)	128
5.4	Agarose gel electrophoresis for PCR product	137
5.5	Conventional PCR machine (Thermo cycler peq start)	138
5.6	Gel electrophoresis of RFLP	142
5.7	Gel doc (Lab India)	143
6.1a	Correlation between HOMA IR and WHR ratio in Group II	151
6.1b	Correlation between HOMA IR and WHR ratio in Group III	151
6.1c	Correlation between HOMA IR and WHR ratio in Group IV	152
6.1d	Correlation between HOMA IR and BMI in Group II	152
6.1e	Correlation between HOMA IR and BMI in Group III	153
6.1f	Correlation between HOMA IR and BMI in Group IV	153
6.3a	Correlation between HOMA IR and TG/HDL ratio in Group II	168
6.3b	Correlation between HOMA IR and TG/HDL ratio in Group III	168
6.3c	Correlation between HOMA IR and TG/HDL ratio in Group IV	169

List of Figures

Figure No.	Name of Figure	Page No.
6.3d	Correlation between HOMA IR and serum Apo B in Group II	169
6.3e	Correlation between HOMA IR and serum Apo B in Group III	170
6.3f	Correlation between HOMA IR and serum Apo B in Group IV	170
6.4a	Correlation between HOMA IR and serum Nitric Oxide in Group II	179
6.4b	Correlation between HOMA IR and serum Nitric Oxide in Group III	179
6.4c	Correlation between HOMA IR and serum Nitric Oxide in Group IV	180
6.4d	Correlation between HOMA IR and urinary microalbumin levels in Group II	180
6.4e	Correlation between HOMA IR and urinary microalbumin levels in Group III	181
6.4f	Correlation between HOMA IR and urinary microalbumin levels in Group IV	181
6.5a	Gel electrophoresis of RFLP	186

List of Graphs

Graph No.	Name of Graphs	Page No.
6.1	Anthropometric Parameters of all Study Groups	147
6.2a	Glycemic parameters (FPG and PPPG) of all Study Groups s	157
6.2b	Glycemic parameters (HbA1c, Plasma Insulin and HOMA IR) in all Study Groups	158
6.3a	Lipid Parameters (TC, TG, HDL) of all Groups	163
6.3b	Lipid Parameters (VLDL, LDL, Apo B) of all Groups	164
6.4a	Endothelial parameters (Serum Nitric oxide) of all Groups	174
6.4b	Endothelial parameters (Urine Microalbumin) of all Groups	175
6.5a	Frequency distribution of eNOS gene polymorphism in Group I	189
6.5b	Frequency distribution of eNOS gene polymorphism in Group II	189
6.5c	Frequency distribution of eNOS gene polymorphism in Group III	190
6.5d	Frequency distribution of eNOS gene polymorphism in Group IV	190

Abbreviations

MS	: Metabolic syndrome
DM	: Diabetes mellitus
T2DM	: Type 2 diabetes
NO	: Nitric oxide
MODY	: Maturity onset diabetes of young
HNF	: Hepatocyte Nuclear factor
IDDM	: Insulin-dependent diabetes mellitus
NIDDM	: Noninsulin-dependent diabetes mellitus
GDM	: Gestational diabetes mellitus
CDC	: Centers for Disease Control and Prevention
WHO	: World Health Organization
IDF	: International Diabetes Federation
PACAP	: Pituitary adenylate cyclase-activating polypeptide
GLP 1	: Glucagon-like peptide-1
GIP	: Glucose-dependent insulinotropic polypeptide
G6P	: Glucose-6-phosphate
ATP	: Adenosine triphosphate
AMP	: Adenosine monophospahte
FPG	: Fasting plasma glocose
PPPG	: Post prandial plasma sugar
ADA	: American diabetes association
CDC	: Centers for disease control
IGT	: Impaired glucose tolerence
IFG	: Impaired fasting glucose
LDL	: Low density lipoprotein
VLDL	: Very Low density lipoprotein
RER	: Rough endoplasmic reticulum
PACAP	: Pituarty adenylate cyclase activating polypeptide
GIP	: Glucose dependent insulintropic polypepetide
GLP-1	: Glucagon like peptide-1
G6P	: Glucose-6-phosphate
VIP	: Vasoactive intestinal peptide
cAMP	: Cyclic adenosine monophospahte
PKC	: Protein Kinase C
NEFA	: Non-esterified fatty acids
GSIS	: Glucose stimulated insulin secretion
IRS	: Insulin responsive substrates
GLUT	: Glucose transporter proteins
IGF	: Insulin like growth factor
CSF	: Cerebrospinal fluid
GOD	: Glucose oxidase

GOD : Peroxidase
HPLC : High performance liquid chromatography

Abbreviations

HDL : High density lipoproteins
OGTT : Oral glucose tolerance test
ELISA : Enzyme linked immunosorbent assay
SMBG : Self-monitoring of blood glucose
CVD : Cardiovascular disease
FFAs : Free fatty acids
EGIR : European Group for the Study of Insulin Resistance
NCEP : National Cholesterol Education Program's Adult Treatment Panel III
IDF : International Diabetes Federation
AACE : American Association of Clinical Endocrinologists
HOMA : Homeostasis Model of Assessment
IR : Insulin resistance
QUICKI : Quantified insulin sensitivity check index
VEGF : Vascular endothelial growth factor
ET-1 : Endothelin-1
vWF : Von Willebrand factor
VCAM-1 : Vascular cell adhesion molecule- 1
MCP-1 : Monocyte chemotactic protein-1
NF-κB : Nuclear factor κB
FMN : Flavin mononucleotide
FMD : Flavin dinucleotide
CaM : Calmodulin
H4B : Tetrahydrobiopterin
NOS : Nitric oxide synthase
eNOS : Endothelial Nitric oxide synthase
WHR : Waist to hip ratio
BMI : Body mass index
Apo B : Apolipoprotein B
TG : Triglycerides
WC : Waist circumference
AGEs : Advanced glycation end products
iNOS : Induced nitric oxide synthase
nNOS : Neuronal Nitric oxide synthase
LPL : Lipoprotein lipase
CEPT : Cholesteryl ester transfer protein
L-NAME : NG-nitro-L-arginine methyl ester
NADPH : Nicotinamide dihydrogen phosphate

Chapter 1

Introduction

Introduction

The Indian subcontinent is undergoing epidemiological transition as non-communicable diseases like coronary heart disease and type 2 diabetes mellitus are fast replacing infections as the leading cause of morbidity and mortality.[1] According to International diabetic federation, the incidence of diabetes in 2011 was 366 million and is expected to rise to 552 million by 2030. The number of people with type 2 diabetes is increasing in every country.[2] India is already the second "Diabetes Capital" of the world[3] and by the year 2020, it is projected to have the highest number of individuals suffering from atherothrombotic cardiovascular disease due to diabetes.[1]

Thus, diabetes mellitus (DM) is a serious worldwide public health problem due to its frequency, chronic complications and their associated high cost of treatment. Type 2 diabetes mellitus (T2DM), formerly non-insulin-dependent diabetes mellitus (NIDDM) or adult-onset diabetes; is a metabolic disorder that is characterized by high blood glucose accompanied in the context of insulin resistance and relative insulin deficiency. It is characterized by multifactorial pathology that involves insulin resistance, dyslipidemia, endothelial dysfunction, inflammation and hypertension.[4]

People with diabetes have an increased risk of developing a number of serious health problems. Consistently high blood glucose levels can lead to serious diseases affecting the heart and blood vessels, eyes, kidneys and nerves. In addition, people with diabetes also have a higher risk of developing infections.[2] In almost all high-income countries, diabetes is a leading cause of retinopathy, nephropathy, neuropathy, cardiovascular diseases & peripheral vascular disease. The contributing factors for diabetes mellitus are: age, lifestyle, physical inactivity, genetic factors, obesity & insulin resistance. It is observed that prevalence of diabetes is very high in individuals suffering from metabolic syndrome (MS).[5]

Metabolic syndrome (MS) is a clinical entity comprising of heterogeneous traits represented by the co-occurrence of abdominal fat, impaired glucose tolerance, dyslipidemia (high triglycerides and low high-density lipoprotein [HDL] cholesterol levels) and hypertension. The pathogenesis of MS is based on insulin resistance, to which a number of factors contribute such as sedentary life, diet, and genetic factors.[6]

In 1998, the American Diabetes Association proposed that MS include: glucose intolerance, central obesity, dyslipidemia (including: hypertriglyceridemia, low HDL-cholesterol, and high LDLcholesterol), hypertension, increase in prothrombotic and anti fibrinolytic risk factors for atherosclerosis, but no specific definition or limits for these disorders were proposed.[7] In 1999, the World Health Organization (WHO) defined the components of MS and established the limits of these components[8] and in 2003, the U.S. National Cholesterol Education Program (NCEP) redefined MS in order to simplify its clinical application and improve diagnosis.[9]

The feature of the MS may persist for many years before the development of full blown condition of diabetes mellitus. Each component of the syndrome conveys increased cardiovascular risk, but as a combination they become much more powerful.[6] Thus while MS may represent one end of the condition the other extreme may be considered as diabetes mellitus with complications.[10]

Though diabetes mellitus is a condition not confined to any one particular organ, the vasculature to different organs is affected due to hyperglycemia. This is reflected as damage to the endothelium. Vascular endothelium is a monocellular structure that lines the entire vascular tree. It is not just an anatomical barrier between circulating blood and vascular muscle cells, but it plays an essential role in a multitude of fundamental physiological pathways: homeostasis, transportation of metabolites between the blood and the tissues, angiogenesis phenomena, repair of vascular lesions and control of muscle tone. The endothelial cells synthesize different vasodilator & constrictor factors in response to physiological stimuli to maintain tissue perfusion. The most important vasodilator is nitric oxide synthase derived nitric oxide (NO).[11]

NO is produced through the conversion of the amino acid L-arginine to L-citrulline by the enzyme nitric oxide synthase (NOS). There are several isoforms: NOS1 isolated from the brain, NOS2 or iNOS, produced by macrophages and NOS3 or eNOS from endothelial cells. eNOS is activated by the pulsatile flow of blood through vessels. eNOS produces NO which diffuses to the vascular smooth muscle (VSM) where it activates the enzyme guanylate cyclase which in turn increases local cyclic GMP concentration and thereby induces relaxation of the VSM. In this way it maintains the diameter of the blood vessel ensuring optimal perfusion of tissues. Endothelial dysfunction is associated with decreased NO availability, either through loss of NO production or through loss of NO biological activity.[11]

Apart from all mentioned factors, genetic factors also play a major role in onset of this disease. Among the genes thus investigated, the endothelial nitric oxide synthase (eNOS) gene has drawn considerable attention because of its substantial contribution to production of NO required for maintenance of health of endothelium. Polymorphism in eNOS gene is implicated in vascular disease associated with diabetes.[5]

Hence the study was undertaken to study the entire spectrum of the disease by assessing endothelial dysfunction in pre diabetic phase i.e metabolic syndrome as well as in diabetic condition with & without complications.

Chapter 2

Review of literature

Part –I

Diabetes Mellitus

Diabetes Mellitus

Diabetes mellitus (DM) refers to a group of common metabolic disorders that share the phenotype of hyperglycemia. Several distinct types of DM are caused by a complex interaction of genetics and environmental factors.[12]

2.1 Types of Diabetes Mellitus

a)Type 1 Diabetes (Immune-mediated diabetes)[13]

This form of diabetes, which accounts for only 5–10% of those with diabetes, previously encompassed by the terms insulin dependent diabetes, type I diabetes, or juvenile-onset diabetes, results from a cellular-mediated autoimmune destruction of the β -cells of the pancreas.[13]

b) Type 2 Diabetes Mellitus (T2DM)

This is a heterogeneous group of disorders characterized by variable degrees of insulin resistance, impaired insulin secretion and increased glucose production. Distinct genetic and metabolic defects in insulin action and/ or secretion give rise to the common phenotype of hyperglycemia in type 2 DM.

2.1.1 Classification of Type 2 Diabetes

DM is classified on the basis of the pathogenic process that leads to hyperglycemia, as opposed to earlier criteria such as age of onset or type of therapy (Figure 2.1). The two broad categories of DM are designated type 1 and type 2. Both types of diabetes are preceded by a phase of abnormal glucose homeostasis as the pathogenic processes progress.

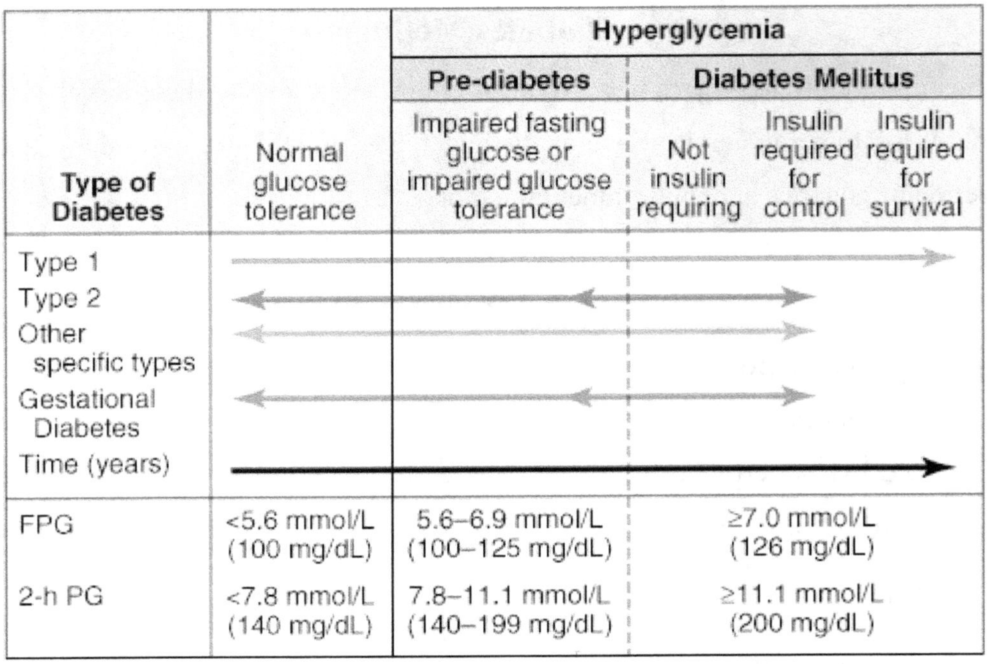

Source: Fauci AS, Kasper DL, Braunwald E, Hauser SL, Longo DL, Jameson JL, Loscalzo J: *Harrison's Principles of Internal Medicine*, 17th Edition: http://www.accessmedicine.com

Figure 2.1 Spectrum of glucose homeostasis and diabetes mellitus (DM). The spectrum from normal glucose tolerance to diabetes in type 1 DM, type 2 DM, other specific types of diabetes, and gestational DM is shown from left to right. In most types of DM, the individual traverses from normal glucose tolerance to impaired glucose tolerance to overt diabetes (these should be viewed not as abrupt categories but as a spectrum). Arrows indicate that changes in glucose tolerance may be bidirectional in some types of diabetes. For example, individuals with type 2 DM may return to the impaired glucose tolerance category with weight loss; in gestational DM, diabetes may revert to impaired glucose tolerance or even normal glucose tolerance after delivery. The fasting plasma glucose (FPG), the 2-h plasma glucose (PG) after a glucose challenge, and the A1C for the different categories of glucose tolerance are shown at the lower part of the figure. These values do not apply to the diagnosis of gestational DM. The World Health Organization uses an FPG of 110–125 mg/dL for the prediabetes category. Some types of DM may or may not require insulin for survival. Some use the term "increased risk for diabetes" (ADA) or "intermediate hyperglycemia" (WHO) rather than "prediabetes," adapted from the American Diabetes Association, 2007.[12]

2.1.2 Etiologic Classification of Diabetes Mellitus[12]

I. Type 1 diabetes (β-cell destruction, usually leading to absolute insulin deficiency)
A. Immune-mediated B. Idiopathic II. Type 2 diabetes (may range from predominantly insulin resistance with relative insulin deficiency to a predominantly insulin secretory defect with insulin resistance)
III. Other specific types of diabetes
A. Genetic defects of cell function characterized by mutations in: 1. Hepatocyte nuclear transcription factor (HNF) 4 (MODY 1) 2. Glucokinase (MODY 2) 3. HNF-1A (MODY 3) 4. Insulin promoter factor-1 (IPF-1; MODY 4) 5. HNF-1B (MODY 5) 6. NeuroD1 (MODY 6) 7. Mitochondrial DNA 8. Subunits of ATP-sensitive potassium channel 9. Proinsulin or insulin conversion
B. Genetic defects in insulin action 1. Type A insulin resistance 2. Leprechaunism 3. Rabson-Mendenhall syndrome 4. Lipodystrophy syndromes
C. Diseases of the exocrine pancreas—pancreatitis, pancreatectomy, neoplasia, cystic fibrosis, hemochromatosis, fibrocalculous pancreatopathy, mutations in carboxyl ester lipase
D. Endocrinopathies-acromegaly, Cushing's syndrome, glucagonoma, pheochromocytoma, hyperthyroidism, somatostatinoma, aldosteronoma

E. Drug or chemical-induced-Vacor, pentamidine, nicotinic acid, lucocorticoids, thyroid hormone, diazoxide, -adrenergic agonists, thiazides,phenytoin, interferon, protease inhibitors, clozapine
F. Infections-congenital rubella, cytomegalovirus, coxsackie
G. Uncommon forms of immune-mediated diabetes-"stiff-person" syndrome, anti-insulin receptor antibodies
H. Other genetic syndromes sometimes associated with diabetes-Down's syndrome, Klinefelter's syndrome, Turner's syndrome, Wolfram's syndrome, Friedreich's ataxia, Huntington's chorea, Laurence-Moon-Biedl syndrome, myotonic dystrophy, porphyria, Prader-Willi syndrome.
IV. Gestational diabetes mellitus (GDM)

Note: MODY, maturity onset of diabetes of the young.

Source: Adapted from American Diabetes Association, 2007.

Two features of the current classification of DM diverge from previous classifications. First, the terms insulin-dependent diabetes mellitus (IDDM) and noninsulin-dependent diabetes mellitus (NIDDM) are obsolete. Since many individuals with T2DM eventually require insulin treatment for control of glycemia, the use of the term NIDDM generated considerable confusion. A second difference is that age is not a criterion in the classification system. Although T1DM most commonly develops before the age of 30, an autoimmune beta cell destructive process can develop at any age. It is estimated that between 5 and 10% of individuals who develop DM after age 30 have type 1 DM. Likewise, T2DM more typically develops with increasing age but is now being diagnosed more frequently in children and young adults, particularly in obese adolescents.

2.1.3 Other Types of diabetes

Other etiologies for DM include specific genetic defects in insulin secretion or action, metabolic abnormalities that impair insulin secretion, mitochondrial abnormalities and a host of conditions that impair glucose tolerance.

Maturity onset diabetes of the young (MODY) is a subtype of DM characterized by autosomal dominant inheritance, early onset of hyperglycemia (usually <25 years). Different type of MODY identified are MODY1, MODY2, MODY3, MODY4 MODY5, MODY6, MODY7, MODY8, MODY9 and impairment in insulin secretion (discussed below). Mutations in the insulin receptor cause a group of rare disorders characterized by severe insulin resistance. DM can result from pancreatic exocrine disease when the majority of pancreatic islets are destroyed. Hormones that antagonize insulin action can also lead to DM. Thus, DM is often a feature of endocrinopathies such as acromegaly and Cushing's disease. Viral infections have been implicated in pancreatic islet destruction but are an extremely rare cause of DM. A form of acute onset of type 1 diabetes, termed fulminant diabetes, has been noted in Japan and may be related to viral infection of islets.

2.1.4 Gestational Diabetes Mellitus (GDM)

Glucose intolerance may develop during pregnancy. Insulin resistance is related to the metabolic changes of late pregnancy and the increased insulin requirements may lead to IGT. GDM occurs in ~4% of pregnancies in the United States; most women revert to normal glucose tolerance post-partum but have a substantial risk (30–60%) of developing DM later in life.

2.2. Epidemiology of diabetes mellitus

The prevalence of type 2 diabetes is increasing rapidly not at the national level but globally also.

2.2.1 Global prevalence of diabetes mellitus

The worldwide prevalence of DM has risen dramatically over the past two decades, from an estimated 30 million cases in 1985 to 177 million in 2000. Based on current trends, >360 million individuals will have diabetes by the year 2030 (Figure 2.2). Although the prevalence of both type 1 and type 2 DM is increasing worldwide, the prevalence of T2DM is rising much more rapidly because of increasing obesity and reduced activity levels as countries become more industrialized. This is true in most countries and 6 of the top 10 countries with the highest rates are in Asia. In the United States, the Centers for

Disease Control and Prevention (CDC) estimated that 20.8 million persons, or 7% of the population, had diabetes in 2005 (~30% of individuals with diabetes were undiagnosed). Approximately 1.5 million individuals (>20 years) were newly diagnosed with diabetes in 2005. DM increases with aging. In 2005, the prevalence of DM in the United Sates was estimated to be 0.22% in those <20 years and 9.6% in those >20 years. In individuals >60 years, the prevalence of DM was 20.9%. The prevalence is similar in men and women throughout most age ranges (10.5% and 8.8% in individuals >20 years) but is slightly greater in men >60 years. Worldwide estimates projects that in 2030 the greatest number of individuals with diabetes will be 45–64 years of age.[12]

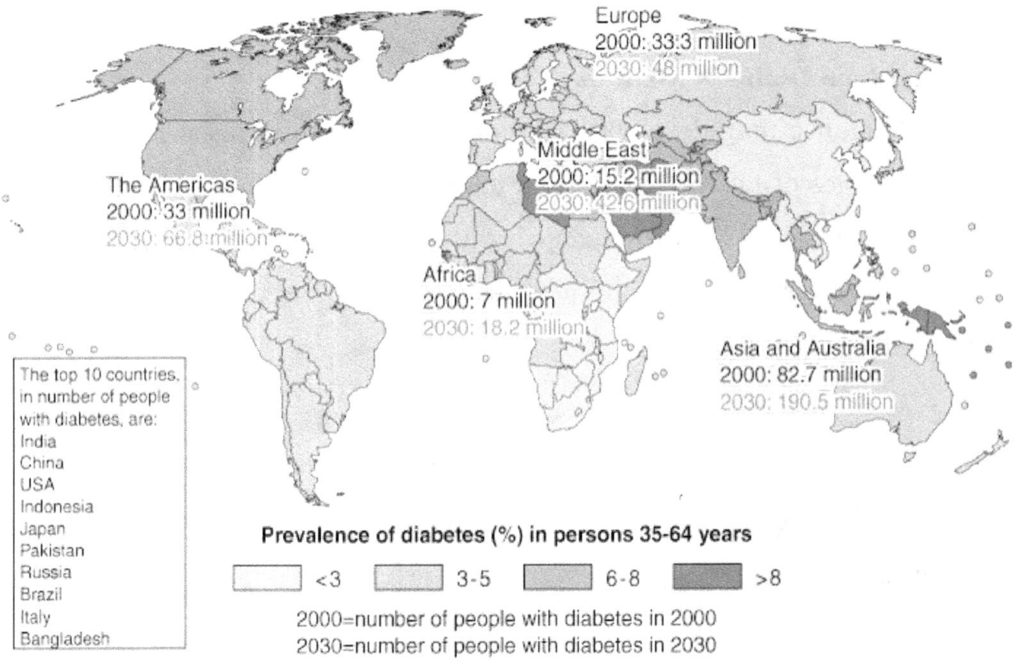

Source: Fauci AS, Kasper DL, Braunwald E, Hauser SL, Longo DL, Jameson JL, Loscalzo J: *Harrison's Principles of Internal Medicine*, 17th Edition: http://www.accessmedicine.com

Figure 2.2 Worldwide prevalence of diabetes mellitus. The prevalence of diabetes in 2000 and the projected prevalence in 2030 are shown by geographical region.[14]

The global increase in the prevalence of diabetes is due to population growth, aging, urbanisation and an increase of obesity and physical inactivity. The primary determinants of the epidemic are the rapid epidemiological transition associated with changes in dietary patterns and decreased physical activity. Unlike in the West, where older populations are most affected, the burden of diabetes in Asian countries is disproportionately high in young to middle-aged adults.[15,16] As per Intenationl Diabetes Federation 2013, some 382 million people worldwide or 8.3% adults were estimated to have diabetes. If these trends continue, by 2035, 592 million people or one adult in 10 will have diabetes. This equates to approximately three new cases every 10 seconds or almost 10 million per year. This could have long-lasting adverse effects on a nation's health and economy, especially for developing countries. Healthcare expenditures on diabetes accounted for 11% of the total healthcare expenditure in the world in 2013.[17] The "Top 10" countries in the world, in terms of the number of people with diabetes, for 2013 and 2035, are shown in Table 2.1 At both time points, the three countries with the largest number of people with diabetes are India, China and the U.S.1 This picture is likely to change soon, in light of the recent escalation in prevalence of diabetes in China.[18] Roughly 80% of people with diabetes are in developing countries, of which India and China share the larger contribution.[15] According to the World Health Organization (WHO) criteria, the prevalence of known diabetes was 5.6% and 2.7% among urban and rural areas, respectively.[19] Ramachandran et al. reported that age-standardised prevalence of diabetes and impaired glucose tolerance (IGT) in urban India in 2000 were 12.1% and 14.0%, respectively, with no gender difference.[20]

Table 2.1: Top 10 countries for estimated numbers of adults with diabetes, 2013 and 2035.[17]

Rank	Country /Territory	2013(millions)	Country /Territory	2035 (millions)
1.	China	98.4	China	142.7
2.	India	65.1	India	109
3.	US	24.4	US	29.7
4.	Brazil	11.9	Brazil	19.2
5.	Russian Federation	10.9	Mexico	15.7
6.	Mexico	8.7	Indonesia	14.1
7.	Indonesia	8.5	Egypt	13.1
8.	Germany	7.6	Pakistan	12.8
9.	Egypt	7.5	Turkey	11.8
10.	Japan	7.2	Russian Federation	11.2

Thus type 2 diabetes, accounts for more than 90 per cent of all diabetes cases. Diabetes epidemic is more pronounced in India as the World Health Organization (WHO) reports showed that 32 million people had diabetes in the year 2002. The International Diabetes Federation (IDF) estimated the total number of diabetic subjects to be around 40.9 million in India and reported that this is further set to rise to 69.9 million by the year 2025. [21]

2.2.2 Evolution of the diabetes epidemic in India

The first national study on the prevalence of type 2 diabetes in India was done between 1972 and 1975 by the Indian Council of Medical Research (ICMR, New Delhi).[22] Screening was done in about 35,000 individuals above 14 yr of age, using 50 g glucose load. Capillary blood glucose level >170 mg/dl was used to diagnose diabetes. The prevalence was 2.1% in urban population and 1.5 % in the rural population while in those above 40 yr of age, the prevalence was 5 % in urban and 2.8 %in rural areas. A national rural diabetes survey was done between 1989 and 1991 in different parts of the country in selected rural populations.[23] This study which used the 1985 WHO criteria to diagnose diabetes, reported a crude prevalence of 2.8 %.[23] A study done in 1988 in Chennai reported a prevalence of 8.2 % in the urban and 2.4 % in the rural areas.[24]A subsequent study in the same urban area done after five years showed an age standardized prevalence of 11.6 % indicating a rising trend in prevalence of diabetes.[25], A very high prevalence of 16.3 % was reported in Thiruvanathapuram in Kerala State in the year 1999.[23] In the same year, a prevalence of 8.2% was reported from Guwahati. [26] A cross-sectional population survey was done in the Kashmir valley in 2000 and the prevalence of 'known diabetes' among adults aged >40 yr was found to be 1.9 %.[27] The National Urban Diabetes Survey (NUDS), a population based study was conducted in six metropolitan cities across India and recruited 11,216 subjects aged 20 yr and above representative of all socio-economic strata. [7]An oral glucose tolerance test was done using capillary glucose and diabetes was defined using the WHO criteria, [28] The study reported that the age standardized prevalence of type 2 diabetes was 12.1 %. This study also revealed that the prevalence in the southern part of India to be higher-13.5 % in Chennai, 12.4 % , in Bangalore and 16.6 % Hyderabad; compared to eastern India (Kolkatta), 11.7 %; northern India (New Delhi), 11.6 %; and western India (Mumbai), 9.3 %. The study also suggested that there was a large pool of subjects with impaired glucose tolerance (IGT), 14 % with a high risk of conversion to diabetes. A study done in western India showed age standardized prevalence of 8.6 % in urban population.[29] A more recent study reported a high prevalence (9.3%) in rural Maharashtra,[30] The Amrita Diabetes and Endocrine Population Survey (ADEPS),[31] a community based cross-sectional survey done in urban areas of Ernakulam district in Kerala has revealed a very high prevalence of 19.5 %. Figure 2.3 is a map of India showing the States where population-based studies have been done and it also shows the prevalence of type 2 diabetes reported in different regions of India.

Further support for the rising prevalence of diabetes comes from the results obtained from the Chennai Urban Rural Epidemiology Study (CURES).[32] The overall crude prevalence of diabetes using WHO criteria[28] in CURES was 15.5 % (age standardized:

14.3%), while that of IGT was 10.6 % (age-standardized: 10.2%). From 1989 to 1995, the prevalence of diabetes in Chennai increased by 39.8 % (8.3 to 11.6%); between 1995 to 2000 by 16.3 % (11.6 to 13.5%) and between 2000 to 2004, by 6.0 % (13.5 to 14.3%). Thus within a span of 14 yr, the prevalence of diabetes increased significantly by 72.3 % (P<0.001).

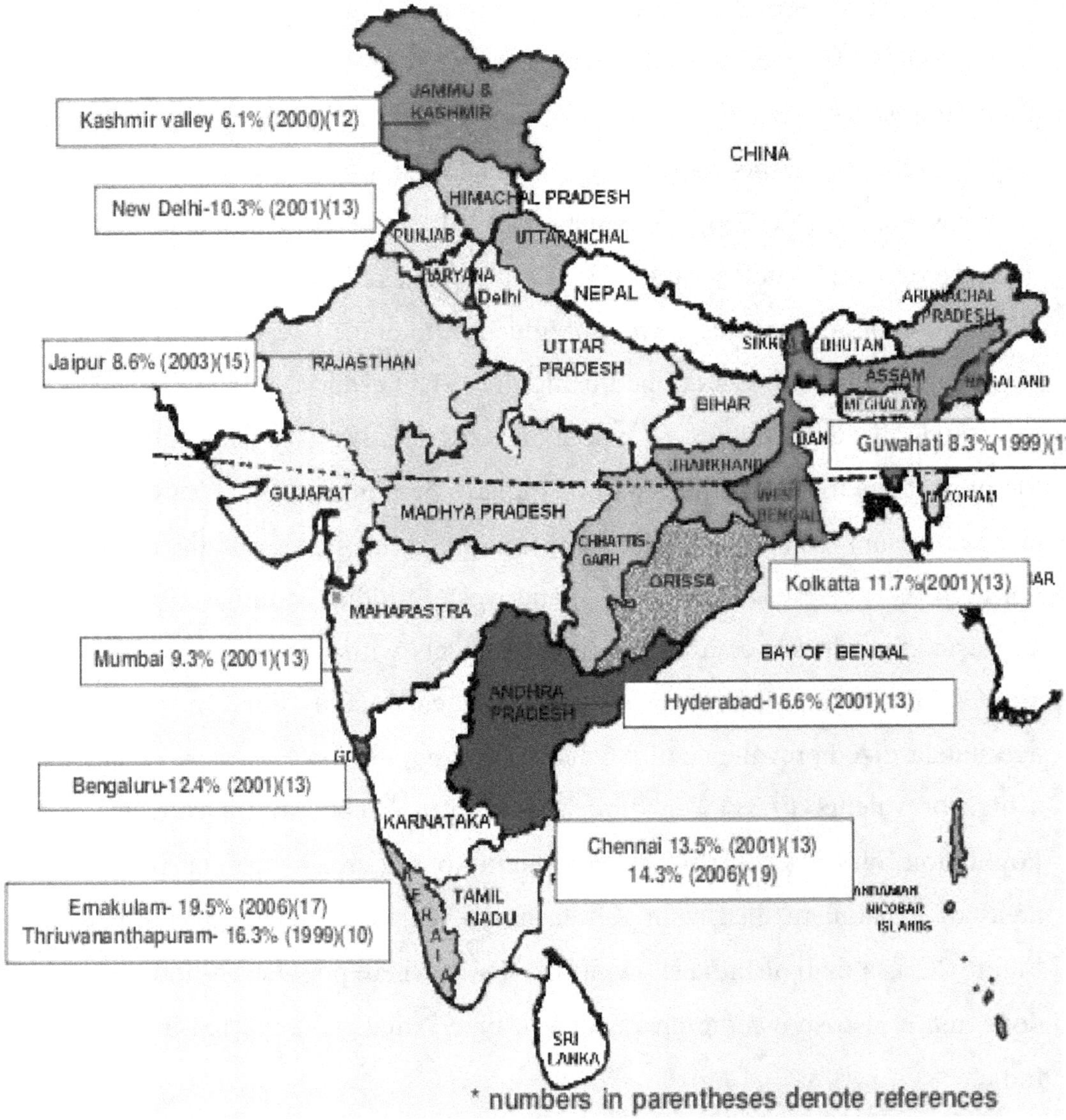

Figure 2.3 Recent population based studies showings the prevalence of type 2 diabetes in different parts of India. [7, 26, 27, 29, 32,33]

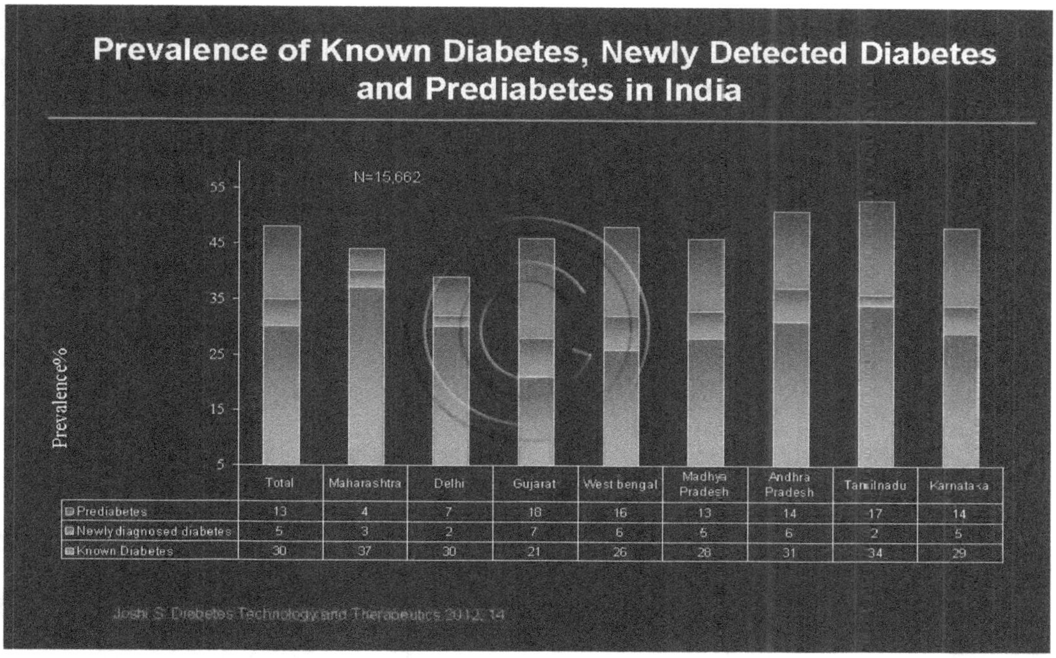

	Total	Maharashtra	Delhi	Gujarat	West bengal	Madhya Pradesh	Andhra Pradesh	Tamilnadu	Karnataka
Prediabetes	13	4	7	18	16	13	14	17	14
Newly diagnosed diabetes	5	3	2	7	6	5	6	2	5
Known Diabetes	30	37	30	21	26	28	31	34	29

Figure 2.4 Prevalence of diagnosed and undiagnosed diabetes and hypertension in India—Results from the Screening India's Twin Epidemic (SITE) Study. [34]

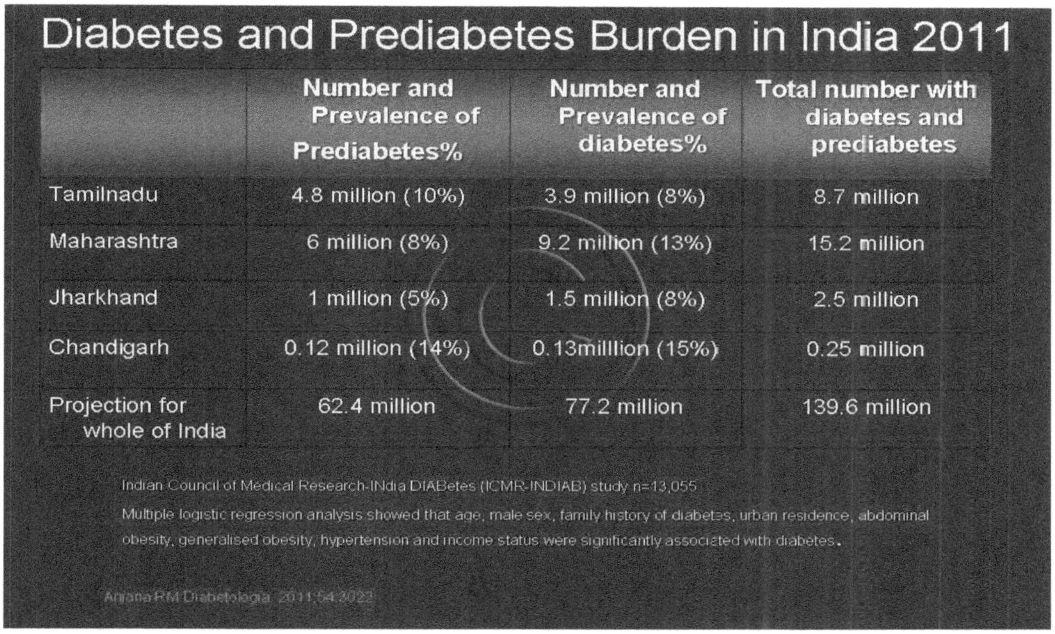

Diabetes and Prediabetes Burden in India 2011

	Number and Prevalence of Prediabetes%	Number and Prevalence of diabetes%	Total number with diabetes and prediabetes
Tamilnadu	4.8 million (10%)	3.9 million (8%)	8.7 million
Maharashtra	6 million (8%)	9.2 million (13%)	15.2 million
Jharkhand	1 million (5%)	1.5 million (8%)	2.5 million
Chandigarh	0.12 million (14%)	0.13milllion (15%)	0.25 million
Projection for whole of India	62.4 million	77.2 million	139.6 million

Indian Council of Medical Research-INdia DIABetes (ICMR-INDIAB) study n=13,055

Multiple logistic regression analysis showed that age, male sex, family history of diabetes, urban residence, abdominal obesity, generalised obesity, hypertension and income status were significantly associated with diabetes.

Anjana RM Diabetologia 2011;54:3022

Figure 2.5 Prevalence of diabetes and prediabetes (impaired fasting glucose and/or impaired glucose tolerance) in urban and rural India: phase I results of the Indian Council of Medical Research-India Diabetes (ICMR-INDIAB) study.[35]

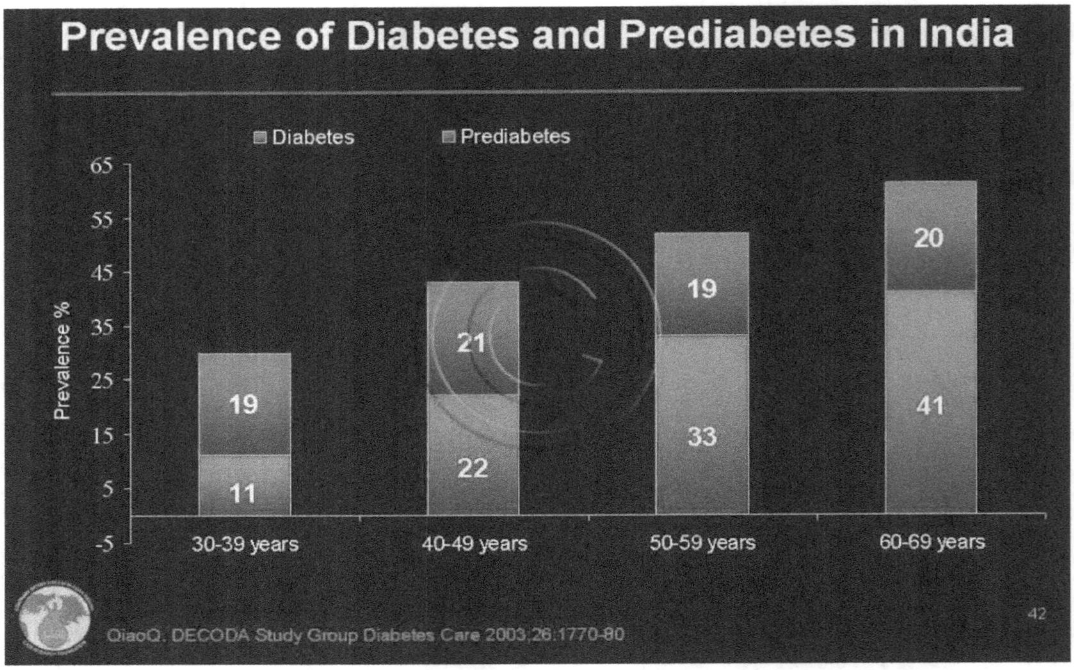

Figure 2.6 Age- and sex-specific prevalence of diabetes and impaired glucose regulation in 11 Asian cohorts[36]

2.3. Historical aspects of Diabetes

Throughout most of its history, diabetes mellitus was considered classified and promulgated as a disease of the kidneys. It was only in the latter part of the 19th century, when chemistry and physiology moved medicine from an observational to an investigative science, that it first came to be defined as a metabolic and shortly thereafter as an endocrine disease.[37]

Susruta and Charaka were the first to discern a difference between two groups of diabetics. They noted the difference between thin people who develop the disease at a young age and heavier people who develop diabetes at an older age. They also noted that the older heavier group seemed to live longer. This method of classifying diabetics remains with us today. We now refer to the first group as type 1 and the second as type 2 diabetes. Susruta and Charuka also re-invented a clinical test for the disease. They noted that the urine of many of the people with polyuria tasted like honey, was sticky and attracted ants when it was poured near an anthill.[38]

2.3.1 History of diabetes in timeline

1st Century A.D.

In the first century AD, the Greeks described the disease as "a melting down of the flesh and limbs into urine."

2nd Century A.D.

The second century A.D. saw significant advancement in the understanding of diabetes. Aretaeus of Cappadocia, a 2nd century A.D. physician, re-labeled the condition "diabetes," and successfully distinguished between diabetes mellitus and diabetes insipidus.

Along with Aretaeus, the roman physician Galen (131 - 201 A.D.), also published his observation about diabetes. Galen encountered only two cases during his entire career. Galen, unlike Aretaeus, labeled the condition diarrhea urinosa and dipsakos, referring to the excessive urine production and thirst experienced by diabetics. Around the same time physicians in China and Japan observed that the urine of diabetics seemed to attract dogs. Li Hsuan and other Chinese physicians of that time also observed the fact that diabetics were prone to infections such as boils and a lung disease resembling tuberculosis. They recommended avoidance of wine, sex and salty cereal products.

865 - 925 A.D.

Arabic medicine excelled during the 9th, 10th and 11th centuries A.D. Arabian writer Al-Rahzi (Rhazea) translated and shared the information of DM from the Hindus. This played a major role in informing the western world of the pioneering work done by Susruta and Charaka.

900 - 1037 A.D.

Among the brilliant contributors to the sciences of Pharmacy and Medicine during the Arabian era was one genius who seems to stand for his time - the Persian, Ibn Sina, called Avicenna by the Western world. He published a monumental medical encyclopedia that accurately described the clinical features of diabetes as well as several of the complications, including gangrene and loss of sexual function.

1000 A.D.

About 1000 AD, Greek physicians prescribed exercise, preferably on horseback, to "employ moderate friction" and alleviate excess urination

125 A.D.

Physicians prescribed use of purgatives to relieve strain on the kidneys, astringents and refrigerant remedies. Von Hohenheim, a Swiss physician, observed a white powder when a diabetic's urine was allowed to evaporate. He concluded, incorrectly, that this residue was salt and therefore that the polyuria observed in diabetes was caused by an excessive thirst of the kidneys resulting from salt deposition in the kidneys.

1501 - 1600 A.D.

Cardona (1501-1576 A.D.) measured the intake and output of fluids, found a great discrepancy in the figures and surmised that people with diabetes lose more water than they take in due to some unknown reason.

1601-1700 A.D.

Thomas Willis (1621-1675), a British physician practicing in the 17th century, re-discovered what Indian physicians had observed a thousand years earlier. The urine of diabetics is sweet-tasting. It had first been identified by the ancient Indians, who referred to diabetes as Madhumeha.The ancient Indians are said to have noticed ants congregating around the urine of diabetics, so perhaps the ants deserve the credit for discovering the sweetness of diabetic urine.

701-1800 A.D

In 1776 British experimental physiologist Matthew Dobson (1735-1784) published the first experimental evidence to demonstrate that diabetes is a systemic disorder rather than, a primary disease of the kidneys. In a series of experiments he discovered a sweet-tasting substance not only in the urine of diabetics but in their serum as well. He went on to prove that this substance was sugar.

In 1788 Thomas Cawley published the first evidence of a link between diabetes and the pancreas. He observed the development of diabetes in people who had sustained injury to the pancreas.

In 1798, another British physician, John Rollo, was the first to add the adjective mellitus to diabetes when he published a paper entitled, "An Account of Two Cases of Diabetes Mellitus." He applied the name mellitus, derived from the Latin and Greek roots for honey, to distinguish diabetes mellitus from other causes of polyuria (excessive urine production) in which the urine has no sweet taste.

1801-1920A.D.

Claude Bernard (1813-1878), one of the 19th century's great diabetes researchers, discovered the fact that the same sugar found in the urine of diabetics is stored in the liver in the form of glycogen. He theorized that DM was caused by glycogenolysis from the glycogen stored in the liver. Bernard's important discoveries were first, that the liver stored glycogen and secreted a sugary substance into the blood. He assumed that it was an excess of this secretion which caused diabetes

L. Traube (1816-1876 A.D.) related that the intake of carbohydrates and its digestion increases the amount of sugar in the urine, that the stopping of the intake of the carbohydrate eliminates most of this sugar in the urine. Paul Langerhans (1847-1888) in 1869, a German medical student, announces in a dissertation that the pancreas contains two systems of cells. One set secretes the normal pancreatic juice; the function of the other was unknown. Several years later, these cells are identified as the "islets of Langerhans." It would be another 30 years before anyone would discover that these islands of cells produce insulin.

Diabetology owes much to the Franco-Prussian War of 1870 and its sequelae. French physician, Bouchardat, notices the disappearance of glycosuria and ketonuria decreased or disappeared in his diabetic patients and so did the cardinal symptoms and signs. These observations supported the view of clinicians who had previously prescribed calorie restriction, periods of fasting and increased muscular work as the rational therapy for the overweight diabetic individual.

In the late 1880's came one of those extraordinary strokes of chance that instantaneously change men's minds, the discovery that removing the pancreas caused diabetes. Von Mering, a German physician in Strasburg, was interested in digestion. He wanted to know what role the pancreas played. He wanted to try the effect of removing it but thought that this was impossible. His colleague Oscar Minkowski said, 'Bah, I can take out the pancreas'. He could and he did. Then the unexpected happened. The next day the laboratory technician complained that the dog was urinating all over the cage, in spite of having been house-trained. Minkowski realised at once what this might mean, tested the urine and found glucose. He had produced diabetes by removing the pancreas, therefore the pancreas contained an antidiabetic substance.

1923A.D.

The Nobel Prize in physiology and medicine is awarded to Frederick Banting and J.J.R. Macleod for their work in extracting insulin from pancreatic tissue. Banting protests the award to Macleod, who had provided laboratory space but had not participated in the discovery, and shares his part of the prize equally with Charles Best.

Eli Lilly and Company begins commercial production of insulin. The insulin was soon available for commercial use. The types of syringes and kits for injections used around this time is quite fascinating.[39]

2.4. Pathophysiology of type 2diabetes

2.4.1 Natural history of type 2 diabetes and Pathophysiology

Type 2 diabetes is chronic progressive, metabolic disease defined by the presence of hyperglycemia. The disease is incompletely understood and represents a complex metabolic condition. It is characterized by hyperglycemia, insulin resistance and decreased β cell function numbers and maximal secretory function, increased glucagon secretion and hepatic glucose production, hypertension, abnormalities in adipocyte and lipid metabolism, rapid gastric emptying, increased appetite, obesity, systemic inflammation, elevated cytokine, hypercoagulation and endothelial cell dysfunction.[40-43]

Underlying the abnormal glucose homeostasis in type 2 diabetes is resistance to some actions of insulin (importantly stimulation of glucose transport and suppression of hepatic glucose release), inadequate secretion of insulin to match metabolic needs and increased production of glucose by the liver. A schematic of the natural history of T2DM is depicted in Figure 2.7. Insulin resistance (the inverse of insulin-sensitivity) is commonly increased in the "pre-diabetes" phase, but normal glucose levels are maintained as well as β cell secret higher amount of insulin.[44]

However, although during the pre-diabetes phase insulin secretion is stimulated, the levels are less than what is needed, this is due less responsiveness of β-cell in glucose and in maximumal β cell secretoy capacity. There is also decreased responsiveness of the liver to insulin (to suppress glucose production by gluconeogenesis and glucose release by glycogenolysis) and increased resistance of muscle to insulin (to stimulate glucose uptake); the resistance to actions of insulin in peripheral tissues, including adipocyte, reflect post receptor defect in insulin signaling pathway. The β-cell, failure which is age related, is mediated by combination of genetic factors, glucose bination of genetic factors, exposure to elevated levels of glucose and free fatty acids in blood (gluco- and lipo-toxicity, respectively) possibly deposition of amyloid fibril in islets and increased demand to secret more insulin in response to ambient hyperglycemia.

It is noteworthy that the bulk of the currently described genetic abnormalities associated with T2DM are related to β-cell function, [45, 46] The critical role of β-cell failure in development of T2DM has significant implication in management of disease.

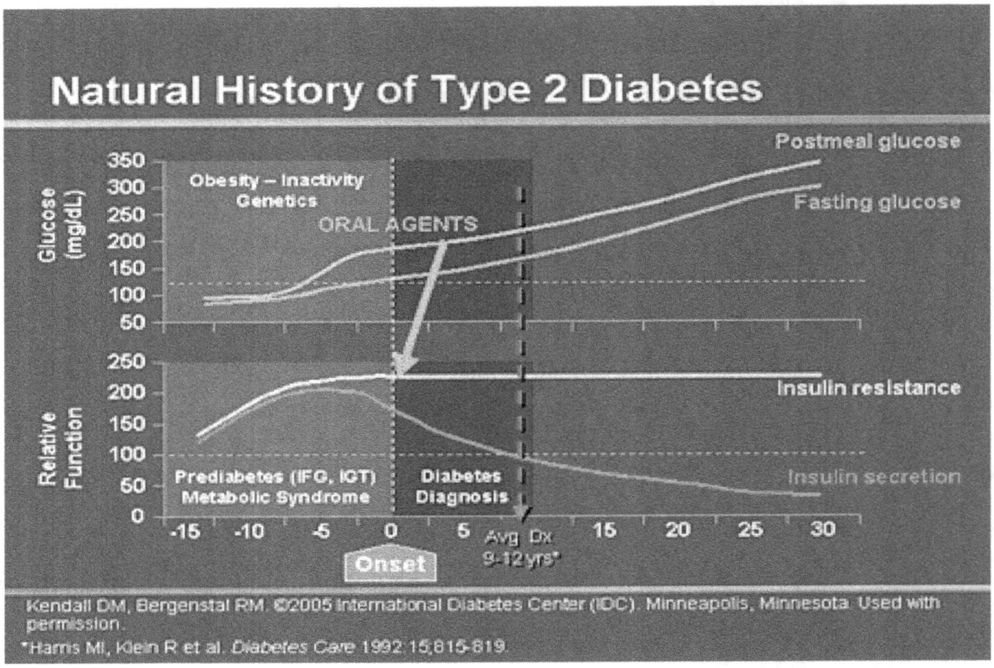

Figure 2.7 Natural history of type 2 diabetes mellitus. The 3 core pathophysiologic defects likely responsible for the progressive nature of the disease and deterioration in glycemic control, insulin resistance, insulin deficiency and impaired incretin effect are highlighted, showing their relative function and course. IFG = impaired fasting glucose; IGT = impaired glucose tolerance.[47]

Examination of figure 2.8 is of great interest. The figure shows the relationship between insulin secretion and insulin sensitivity in normal euglycemic individuals. Critical to our understanding is the fact that there is great variation in insulin secretion as well as in insulin sensitivity among normal individuals. The underlying causes for this vast variation are unclear; most of the variation is not explained by adipocity and probably reflect genetic and epigenetic factors. Although the figure depicts relationship among different normal individuals, each person has the ability of moving left or right on the X-axis (insulin sensitivity). For example, a person who gains weight may have a decrease in his or her insulin sensitivity, but will continue to have normal blood glucose levels as long as insulin secretion increases. Hence, T2DM represents a condition where insulin secretion is much less than what is needed, given the degree of insulin resistance in the individual.

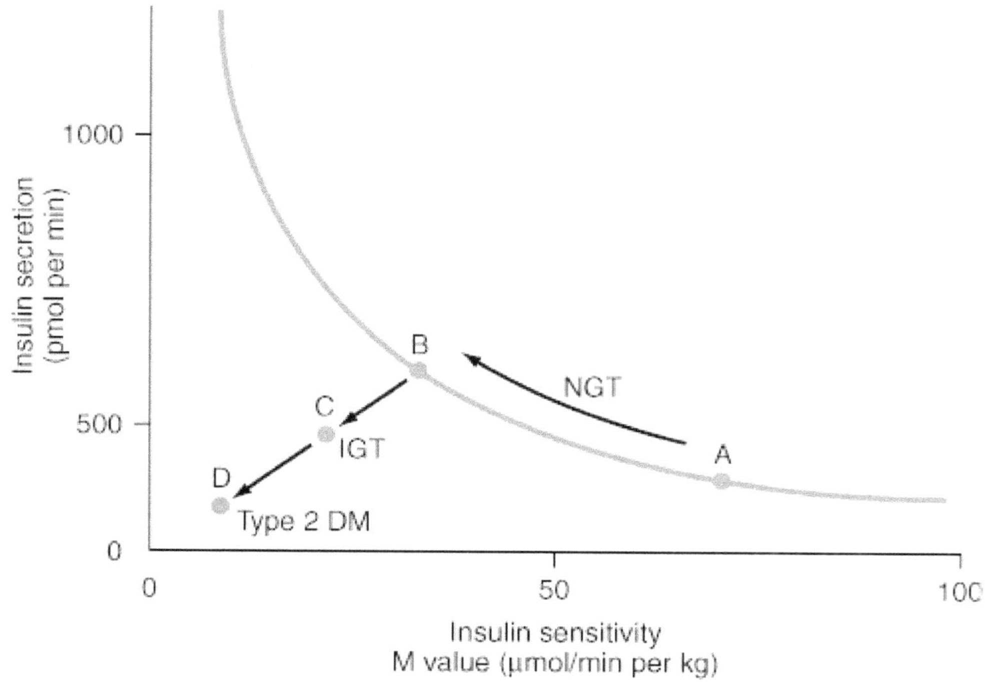

Source: Fauci AS, Kasper DL, Braunwald E, Hauser SL, Longo DL, Jameson JL, Loscalzo J: *Harrison's Principles of Internal Medicine*, 17th Edition: http://www.accessmedicine.com

Figure 2.8 Metabolic changes during the development of type 2 diabetes mellitus (DM). Insulin secretion and insulin sensitivity are related, and as an individual becomes more insulin resistant (by moving from point A to point B), insulin secretion increases. A failure to compensate by increasing the insulin secretion results initially in impaired glucose tolerance (IGT; point C) and ultimately in type 2 DM (point D).[48]

2.5. Metabolic changes during the development of type 2 diabetes mellitus

Insulin secretion and insulin sensitivity which are the common features of type 2 diabetes are related, and as an individual becomes more insulin resistant (by moving from point A to point B; figure 2.8), insulin secretion increases. A failure to compensate by increasing the insulin secretion results initially in impaired glucose tolerance (IGT; point C; figure 2.8) and ultimately in type 2 DM (point D; fig 2.8).[48]

2.5.1 Abnormal muscle and fat metabolism

Insulin resistance, the decreased ability of insulin to act effectively on target tissues (especially muscle, liver, and fat), is a prominent feature of T2DM and results from a combination of genetic susceptibility and obesity. Insulin resistance is relative, however, since supra normal levels of circulating insulin will normalize the plasma glucose. Insulin dose-response curves exhibit a rightward shift, indicating reduced sensitivity and a reduced maximal response, indicating an overall decrease in maximum glucose utilization (30–60 % lower than in normal individuals). Insulin resistance impairs glucose utilization by insulin-sensitive tissues and increases hepatic glucose output; both effects contribute to the hyperglycemia. Increased hepatic glucose output predominantly accounts for increased FPG levels, whereas decreased peripheral glucose usage results in postprandial hyperglycemia. In skeletal muscle, there is a greater impairment in nonoxidative glucose usage (glycogen formation) than in oxidative glucose metabolism through glycolysis. Glucose metabolism in insulin-independent tissues is not altered in T2DM. The precise molecular mechanism leading to insulin resistance in T2DM has not been elucidated. Insulin receptor levels and tyrosine kinase activity in skeletal muscle are reduced, but these alterations are most likely secondary to hyperinsulinemia and are not a primary defect. Therefore, "post receptor" defects in insulin-regulated phosphorylation/dephosphorylation appear to play the predominant role in insulin resistance. For example, a PI-3-kinase signaling defect might reduce translocation of GLUT4 to the plasma membrane. Other abnormalities include the accumulation of lipid within skeletal myocytes which may impair mitochondrial oxidative phosphorylation and reduce insulin-stimulated mitochondrial ATP production. Impaired fatty acid oxidation and lipid accumulation within skeletal myocytes also may generate reactive oxygen species such as lipid peroxides. Of note, not all insulin signal transduction pathways are resistant to the effects of insulin (e.g., those controlling cell growth and differentiation using the mitogenic-activated protein kinase pathway). Consequently, hyperinsulinemia may increase the insulin action through these pathways, potentially accelerating diabetes-related conditions such as atherosclerosis. The obesity accompanying T2DM, particularly in a central or visceral location, is thought to be part of the pathogenic process. The increased adipocyte mass leads to increased levels of circulating free fatty acids and other fat cell products. For example, adipocytes secrete a number of biologic products (nonesterified free fatty acids, retinol-binding protein 4, leptin, TNF-α, resistin, and adiponectin). In addition to regulating body weight, appetite and energy expenditure, adipokines also modulate insulin sensitivity. The increased production of free fatty acids and some adipokines may cause insulin resistance in skeletal muscle and liver. For example, free fatty acids impair glucose utilization in skeletal muscle, promote glucose production by the liver and impair beta cell function. In contrast, the production by adipocytes of adiponectin, an insulin-sensitizing peptide, is reduced in obesity and this may contribute to hepatic insulin resistance.[12]

2.5.2 Impaired insulin secretion

Insulin secretion and sensitivity are interrelated figure 2.8. In type 2 DM, insulin secretion initially increases in response to insulin resistance to maintain normal glucose tolerance. Initially, the insulin secretory defect is mild and selectively involves glucose-stimulated insulin secretion. The response to other nonglucose secretagogues, such as arginine, is preserved. Abnormalities in proinsulin processing are reflected by increased secretion of proinsulin in type 2 diabetes. Eventually, the insulin secretory defect progresses to a state of inadequate insulin secretion. The reason(s) for the decline in insulin secretory capacity in T2DM is unclear. The assumption is that a second genetic defect-superimposed upon insulin resistance-leads to beta cell failure. Beta cell mass is decreased by approximately 50 percent in individuals with longstanding type 2 diabetes. Islet amyloid polypeptide or amylin is cosecreted by the beta cell and forms the amyloid fibrillar deposit found in the islets of individuals with long-standing T2DM. Whether such islet amyloid deposits are a primary or secondary event is not known. The metabolic environment of diabetes may also negatively impact islet function. For example, chronic hyperglycemia paradoxically impairs islet function ("glucose toxicity") and leads to a worsening of hyperglycemia. Improvement in glycemic control is often associated with improved islet function. In addition, elevation of free fatty acid levels ("lipotoxicity") and dietary fat may also worsen islet function.[12]

2.5.3 Increased hepatic glucose and lipid production

In T2DM, insulin resistance in the liver reflects the failure of hyperinsulinemia to suppress gluconeogenesis which results in fasting hyperglycemia and decreased glycogen storage by the liver in the postprandial state. Increased hepatic glucose production occurs early in the course of diabetes, though likely after the onset of insulin secretory abnormalities and insulin resistance in skeletal muscle. As a result of insulin resistance in adipose tissue, lipolysis and free fatty acid flux from adipocytes are increased, leading to increased lipid [very low density lipoprotein (VLDL) and triglyceride] synthesis in hepatocytes. This lipid storage or steatosis in the liver may lead to nonalcoholic fatty liver disease and abnormal liver function tests. This is also responsible for the dyslipidemia found in T2DM [elevated triglycerides, reduced high-density lipoprotein (HDL), and increased small dense low-density lipoprotein (LDL) particles]. [12]

2.6 Glucose homeostasis

Diabetes mellitus is one of the clinical manifestations of long-term metabolic abnormalities involving multiple organs and hormonal pathways that impair the body's ability to maintain glucose homeostasis. As a result of impaired glucose homeostasis is a hyperglycemia. Prolonged elevation of blood glucose concentrations causes a number of complications like blindness, renal failure, cardiac and peripheral vascular disease, neuropathy, foot ulcers and limb amputation. Vascular complications represent the leading cause of mortality and morbidity in diabetic patients.[49]

2.6.1 Definition of glucose homeostasis

Most tissues and organs need glucose constantly, as an important source of energy. The low blood concentrations of glucose can cause seizures, loss of consciousness and death. On the other hand, long lasting elevation of glucose concentrations, can result in blindness, renal failure, vascular disease etc. therefore, blood glucose concentrations need to be maintained within narrow limits. The process of maintaining blood glucose at a steady-state level is called "glucose homeostasis". [49] This is accomplished by the finely hormone regulation of peripheral glucose uptake, hepatic glucose production, and glucose uptake during carbohydrates ingestion.

2.6.2 Mechanisms of glucose homeostasis

To avoid postprandial hypoglycemia and fasting hypoglycemia, the body can adjust glucose levels by secreting two hormones, insulin and glucagon that work in opposition to each other. During periods of hyperglycemia, the β-cells of the pancreatic islets of Langerhans secrete more insulin. Insulin is synthesized in β-cells of pancreas in response to an elevation in blood glucose and amino acid after a meal. The major function of insulin is to counter the concerned action of a number of hyperglycemia-generating hormones to maintain low blood glucose levels. It also plays an important role in the regulation of glucose metabolism. This hormone regulates glucose metabolism at many sites reducing hepatic glucose output, via decreased gluconeogenesis and glycogenolysis, facilitates the transport of glucose into striated muscle and adipose tissue and inhibits glucagon secretion. Insulin is not secreted if the blood concentration is ≤ 3 mmol/L, but is secreted in increasing amounts as glucose concentrations increase beyond this threshold.[50]

2.7. Insulin

Insulin is a peptide hormone secreted by the β cells of the pancreatic islets of Langerhans and maintains normal blood glucose levels by facilitating cellular glucose uptake, regulating carbohydrate, lipid and protein metabolism and promoting cell division and growth through its mitogenic effects.[51]

2.7.1 Discovery of insulin

During the first two decades of the 20th century, several investigators prepared extracts of pancreas that were often successful in lowering blood sugar and reducing glycosuria in test animals. However, they were unable to remove impurities and toxic reactions prevented its use in humans with diabetes. In the spring of 1921, Frederick G. Banting, a young Ontario orthopedic surgeon, was given laboratory space by J.J.R. Macleod, the head of physiology at the University of Toronto, to investigate the function of the pancreatic islets. A student assistant, Charles Best and an allotment of dogs were provided to test Banting's hypothesis that ligation of the pancreatic ducts before extraction of the pancreas, destroys the enzyme-secreting parts, whereas the islets of Langerhans, which were believed to produce an internal secretion regulating sugar metabolism, remained intact. He believed that earlier failures were attributable to the destructive action of trypsin. The name "insuline" had been introduced in 1909 for this hypothetic substance. Their experiments produced an extract of pancreas that reduced the hyperglycemia and glycosuria in dogs made diabetic by the removal of their pancreases. They next developed a procedure for extraction from the entire pancreas without the need for duct ligation. This extract, now made from whole beef pancreas, was successful for treating humans with diabetes. Facilitating their success was a development in clinical chemistry that allowed blood sugar to be frequently and accurately determined in small volumes of blood. Success with purification was largely the work of J.B. Collip. Yield and standardization were improved by cooperation with Eli Lilly and Company. The Nobel Prize was awarded to Banting and Macleod for the discovery of insulin which Banting shared with Best.[52]

Dr. Rosalyn Yalow and her collaborator, Dr. Solomon Berson, received the Nobel Prize in Physiology or Medicine in 1977 for the discovery of the radioimmunoassay, which led to the creation of bigger and better assays and ushered in a revolution in the measurement of hormones.[53]

2.7.2 Structure and Chemical Properties of Insulin

Insulin was found to be a polypeptide in 1928 with its amino acid sequence identified in 1952. It is in fact a dipeptide, containing A and B chains respectively, linked by disulphide bridges and containing 51 amino acids, with a molecular weight of 5802. Its iso-electric point is pH 5.5. [54] The A chain comprises 21 amino acids and the B chain 30 amino acids. The A chain has an N-terminal helix linked to an anti-parallel C terminal helix; the B chain has a central helical segment. The two chains are joined by 2 disulphide bonds, which join the N- and C-terminal helices of the A chain to the central helix of the B chain. In pro-insulin, a connecting peptide links the N-terminus of the A chain to the C-terminus of the B chain.[55]

2.7.3 Synthesis and Release of Insulin

Insulin is coded on the short arm of chromosome 11[7] and synthesised in the β cells of the pancreatic islets of Langherhans as its precursor, proinsulin. Proinsulin is synthesised in the ribosomes of the rough endoplasmic reticulum (RER) from mRNA as pre-proinsulin. Pre-proinsulin is formed by sequential synthesis of a singal peptide, the B chain, the connecting (C) peptide and then the A chain comprising a single chain of 100 amino acids. Removal of the signal peptide forms proinsulin, which acquires its characteristic three dimensional structures in the endoplasmic reticulum. Secretory vesicles transfer proinsulin from the RER to the Golgi apparatus, whose aqueous zinc and calcium rich environment favours formation of soluble zinc-containing proinsulin hexamers.[55] As immature storage vesicles form from the Golgi, enzymes acting outside the Golgi convert proinsulin to insulin and C peptide.[56] Insulin forms zinc-containing hexamers which are insoluble, precipitating as chemically stable crystals at pH 5.5. When mature granules are secreted into the circulation by exocytosis, insulin and an equimolar ratio of C-peptide are released. Proinsulin and zinc typically comprise no more than 6 % of the islet cell secretion.[55] Insulin secretion from the islet cells into the portal veins is characteristically pulsatile, reflecting the summation of coordinate secretory bursts from millions of islet cells. An ultradian oscillatory pattern of insulin release, in addition to post meal variation, has been reported.[57] In response to a stimulus such as glucose, insulin secretion is characteristically biphasic, with an initial rapid phase of insulin secretion, followed by a less intense but more sustained release of the hormone.[58] Given insulin's pivotal role in glucose utilisation and metabolism, it is not surprising that glucose has multiple influences on insulin biosynthesis and secretion. However, other factors such as amino acids, fatty acids, acetylcholine, pituitary adenylate cyclase-activating polypeptide (PACAP), glucose-dependent insulinotropic polypeptide (GIP), glucagon-like peptide-1 (GLP-1), and several other agonists, together in combination, also influence these processes.[58] Mechanisms of Insulin secretion increased levels of glucose induce the "first phase" of glucose-mediated insulin secretion by release of insulin from secretory granules in the β cell. Glucose entry into the β cell is sensed by glucokinase, which phosphorylates glucose to glucose-6-phosphate (G6P), generating ATP.[59] Closure of K+-ATP-dependent channels results in membrane depolarization and activation of voltage dependent calcium channels leading to an increase in intracellular calcium concentration; this triggers pulsatile insulin secretion.[60] Augmentation of this response occurs by both a K^+-ATP channel-independent Ca^{2+}-dependent pathway and K^+-ATP channel-independent Ca^{2+}-independent pathways of glucose action.[58] Other mediators of insulin release include activation of phospholipases and protein kinase C (e.g. by acetycholine) and by stimulation of adenylyl cyclase activity and activation of β cell protein kinase A, which potentiates insulin secretion. This latter mechanism may be activated by hormones, such as vasoactive intestinal peptide (VIP), PACAP, GLP-1 and GIP. These factors appear to play a significant role in the second phase of glucose mediated insulin secretion, after refilling of secretory granules translocated from reserve pools. [58] Regulation and mechanisms of insulin secretion at the cellular level synthesis and secretion of insulin is regulated by both nutrient and non-nutrient secretagogues, in the context of environmental stimuli and the interplay of other hormones.[56] Nutrient secretagogues such as glucose appear to trigger insulin secretion from the β cell by increasing intracellular ATP and closing of K^+-ATP channels as outlined above. Generation of cyclic AMP and other cellular energy intermediates are also augmented, further enhancing insulin release. Glucose does not require insulin action to enter the β cell (nor do fructose, mannose or galactose).[56] Non-nutrient secretagogues may act via neural stimuli such as cholinergic and adrenergic pathways, or through peptide hormones and cationic amino acids.

Neural Stimuli - Cholinergic transmission; it has been well recognised that vagus nerve stimulation results in pancreatic insulin secretion. This is thought to mediate the so-called "cephalic phase" of insulin secretion, occurring when food is seen, smelled or acutely ingested. Islet cell cholinergic muscarinic receptors activate phospholipase C, with subsequent intracellular events activating protein kinase C, phospholipase A2 and mobilizing intracellular calcium. Insulin secretion by these mechanisms does not occur in the fasting state or if blood glucose levels are low, but may augment the anabolic response to feeding.[56]

Catecholamines, through α2-adrenoceptors, typically inhibit insulin release during stress and exercise. The pancreas also has β-adrenoceptors, which enhance insulin secretion via cAMP, but this effect appears to be minor. The α2-mediated effects appear to act downstream to those of nutrient secretogogues.[56]

Peptide Hormones- A number of peptide hormones influence insulin secretion. The gut hormones GLP-114 and somatostatin enhance and inhibit insulin action respectively. Their secretion is stimulated by nutrients in the gut. Their mechanisms of action appear to be, at least in part, mediated by cAMP generation and activation of a cAMP-responsive protein kinase.[56]Leptin and adiponectin appear to act in a similar fashion whereas acylation-stimulating protein appears to work via glucose phosphorylation, calcium influx and protein kinase C.[61]

Amino Acids-The insulinotropic effect of the cationic amino acid arginine is well established. L-Ornithine has similar effects. Arginine's action is associated with an increase in potassium permeability without any effect on proinsulin biosynthesis. Uptake of arginine into the β cell is associated with depolarization of the cell membrane and gating of voltage dependent calcium channels.[56] Leucine, an essential branched chain amino acid, is a potent enhancer of insulin secretion, allosterically stimulating glutamate dehydrogenase and forming α-ketoglutarate via the Krebs cycle to generate ATP.[48] The major pathways and influences on insulin secretion are presented schematically in Figure 2.9.

2.7.4 Factors Influencing β Cell Mass

Longer term regulation of insulin secretion may be mediated via effects on β cell mass. Growth hormone, prolactin, placental lactogen and GLP-1 not only increase glucose stimulated insulin release and insulin gene expression, but also increase β cell proliferation. [58, 62]

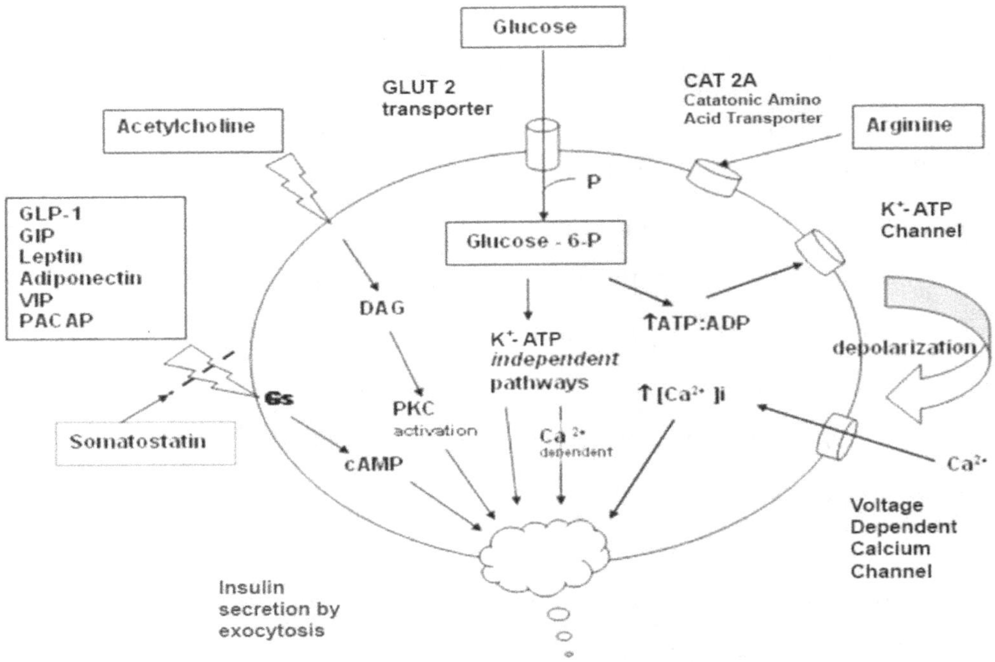

GLP1-Glucagon-like peptide 1 receptor

PACAP-Pituitary adenylate cyclase-activating polypeptide

VIP-Vasoactive intestinal polypeptide

GIP- Gastric inhibitory polypeptide

PKC-Protein Kinase C

Figure 2.9. Schematic presentation of insulin secretory pathways.[58]

2.8 Physiology of Insulin Secretion and Response to various Stimuli

Glucose is the principal stimulus for insulin secretion, though other macronutrients, hormones, humoral factors and neural input may modify this response. Insulin, together with its principal counter-regulatory hormone glucagon, regulates blood glucose concentrations.[63] Pancreatic β cells secrete 0.25-1.5 units of insulin per hour during the fasting (or basal) state, sufficient to enable glucose insulin-dependent entry into cells. This level prevents uncontrolled hydrolysis of triglycerides and limits gluconeogenesis, thereby maintaining normal fasting blood glucose levels. Basal insulin secretion accounts for over fifty percent of total twenty hour insulin secretion. Following secretion of insulin into the portal venous system, sixty percent is subsequently removed by the liver; so portal vein insulin concentrations reaching the liver approach triple that of the peripheral circulation. In healthy lean individuals circulating venous (or arterial) fasting insulin concentrations are about 3-15 mIU/L or 18-90 pmol/L.Meal-related insulin secretion accounts for the remaining fraction of the total daily output. [63]

2.8.1 Response to Glucose

In healthy individuals glucose stimulated pancreatic secretion is biphasic. Intravenous administration of glucose is associated with a rapid "first phase" of insulin release within 1 minute, peaking at 3-5 minutes and lasting about 10 minutes; the slower onset "second phase" of insulin secretion begins shortly after the glucose bolus but is not apparent until 10 minutes later, lasts the duration of the hyperglycaemia and is proportional to the glucose concentration immediately prior to the glucose administration. The first phase of insulin secretion represents release of insulin already synthesized and stored in secretory granules; the second phase represents secretion of both stored and newly synthesised insulin. [63]

2.8.2 Response to Arginine

Arginine induced insulin secretion is proportional to the basal glucose level and occurs 2-10 minutes after IV injection.[63] Oral ingestion of amino acids stimulates the secretion of insulin, as well as growth hormone and glucagon.[64]

2.8.3 Effects of Lipids

In the circulation, non-esterified fatty acids (NEFA) may be derived from dietary lipids and released by the liver, or synthesised in adipocytes, liver, skin and small intestine from excess carbohydrate. NEFA not only increases hepatic glucose output and reduces peripheral insulin sensitivity, but may also modifies glucose stimulated insulin secretion (GSIS). Acute elevation of plasma NEFA is associated with an increase in GSIS, dependent on the prevailing glucose concentration. Chronic elevation of NEFA is associated with a decrease in GSIS and reduced insulin synthesis. Ingestion of lipid may also modify the insulin response to glucose by effects on gastrointestinal hormones and gastric emptying.[65]

2.8.4 Insulin Receptors and Insulin Binding

Insulin mediates its actions through binding to insulin receptors. The insulin receptor was first characterised in 1971. It consists of a heterotetramer consisting of 2 α and 2 β glycoprotein subunits linked by disulphide bonds and is located on the cell membrane.[66] The gene coding for the insulin receptor is located on the short arm of chromosome nineteen.[63]Insulin binds to the extracellular α subunit, resulting in conformational change enabling ATP to bind to the intracellular component of the β subunit. ATP binding in turn triggers phosphorylation of the β subunit conferring tyrosine kinase activity. This enables tyrosine phosphorylation of intracellular substrate proteins known as insulin responsive substrates (IRS). The IRS can then bind other signaling molecules which mediate further cellular actions of insulin.[66] There are four known specifically-named IRS proteins. IRS 1 and 2 have widely overlapping tissue distribution. IRS 1 is phosphorylated by both the insulin receptor and insulin like growth factor 1 receptor, mediates the mitogenic effects of insulin and couples glucose sensing to insulin secretion with IRS 1 proposed to be the major IRS in skeletal muscle. IRS 2, proposed to be the main IRS in liver, mediates peripheral actions of insulin and growth of pancreatic β cells.[66] IRS 3 and 4 are less well characterized. IRS 3 is found only in adipose tissue, β cells and liver and IRS 4 in thymus, brain and kidney.[67,68] Phosphorylated IRS proteins bind specific src-homology-2 domain proteins (SH2) which include important enzymes such as phosphatidylinositol 3-kinase (PI 3-kinase) and phosphotyrosine phosphatase SHPTP2 (or Syp) and other proteins that lack enzymatic activity but which link IRS-1 and other intracellular signaling systems, e.g. the adaptor protein Grb2 which connects with the RAS (rat sarcoma protein) pathway. PI 3-kinase promotes the translocation of glucose transporter proteins, glycogen, lipid and protein synthesis, anti-lipolysis and the control of hepatic gluconeogenesis.[68] PI 3-kinase acts via serine and threonine kinases such as Akt/protein kinase B (PKB), protein kinase C (PKC) and PI dependent protein kinases 1& 2 (PIPD 1&2). The RAS pathway activates transcription factors and stimulates the growth promoting actions of insulin.[66] Thus broadly, PI 3-kinase mediates insulin's metabolic effects, e.g. cellular glucose uptake, while RAS significantly mediates insulin's mitogenic effects [63,66] together with other less well described actions. These pathways are presented schematically in Figure 2.10.

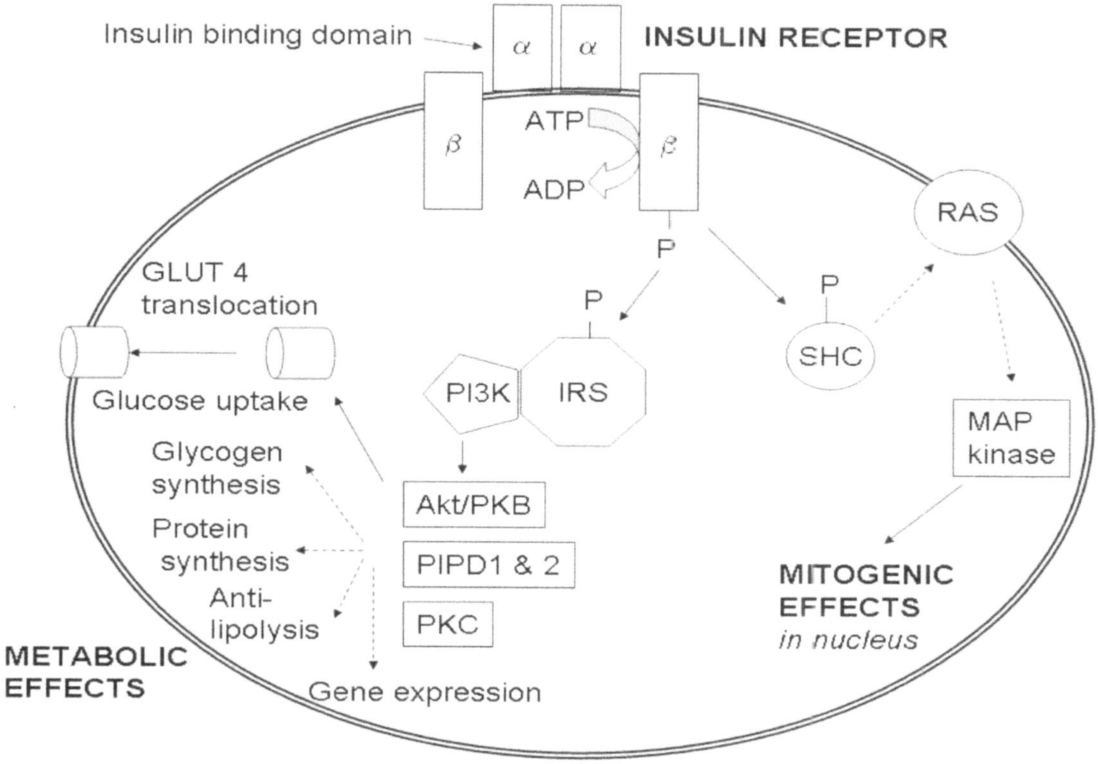

Figure 2.10. Schematic presentation of insulin signalling pathways.

Footnotes: Figure Abbreviations [66,69,70]

Akt/PKB = protein kinase B
ADP = adenosine diphosphate
DAG = diacylglycerol
glucose-dependent insulinotropic polypeptide
GLUT 2 = glucose transport protein 2
GLP-1 = glucagon-like peptide 1
MAP kinase = mitogen activated protein kinase
RAS = rat sarcoma protein
PI3K = phosphatidylinositol 3-kinase

PKC = protein kinase C
vasoactive intestinal peptide

ATP = adenosine triphosphate
cAMP= cyclic adenosine monophosphate
GIP = gastric inhibitory peptide /
Glucose- 6- P = glucose 6 phosphate
GLUT 4 = glucose transport protein 4
IRS = insulin receptor substrate
PACAP = pituitary adenylate cyclase-activating polypeptide
PIPD1 & 2 = phosphatidylinositol dependent protein kinases 1 & 2
SHC =adaptor protein with src-homology VIP=

2.8.5 Glucose Transporter Proteins

Glucose enters cells in an ATP-independent manner by means of glucose transporter proteins (GLUT), of which at least 5 subtypes have been identified. [69,71] Differing in characteristics such as Km for maximal glucose transport and insulin dependency, they enable different cell types to utilise glucose according to their specific functions. For example, most brain cells, having GLUT 1 as the principal transporter protein, are able to move glucose intracellularly at very low blood glucose concentrations without the need for insulin. Thus these neurons, which are principally dependent on glucose for intracellular energy, are able to extract it from the circulation and function despite the low glucose and insulin levels seen during the fasting state. On the other hand, adipose cells and muscle cells have GLUT 4 as the major glucose transporter protein, which requires insulin for its action and has a much higher *Km* for glucose. This enables adipose tissue cells, whose function is to store excess energy, to respond to the higher glucose levels characteristic of the fed state and allows glucose to enter the cells where fatty acid and glycerol synthesis is stimulated and lipolysis suppressed. However, where glucose and insulin levels fall to fasting values, glucose no longer enters the cells, promoting lipolysis. In muscle cells, intracellular glucose transport facilitates glycogen synthesis in the fed state.[68] PI 3-kinase appears to be essential for the translocation of GLUT 4 to the cell membrane in muscle cells and adipocytes; this facilitates the downstream actions of this key intracellular enzyme.[66]

2.8.6. Actions of Insulin on different metabolisms

Insulin's actions at the cellular level encompass carbohydrate, lipid and amino acid metabolism and mRNA transcription and translation.

A. Carbohydrate Metabolism

Insulin acts at multiple steps in carbohydrate metabolism. Glycogen synthesis is increased and glycogen breakdown decreased, by dephosphorylation of glycogen synthase and glycogen phosphorylase kinase respectively. Glycolysis is stimulated and gluconeogenesis inhibited by dephosphorylation of pyruvate kinase (PK) and 2,6 biphosphate kinase. Signalling intracellular energy abundance, insulin enhances the irreversible conversion of pyruvate to Acetyl Co-A by activation of the intra-mitochondrial enzyme complex pyruvate dehydrogenase. Acetyl-CoA may then be directly oxidised via the Krebs' cycle, or used for fatty acid synthesis.[72]

B. Lipid Metabolism

Insulin stimulates fatty acid synthesis in adipose tissue, liver and lactating mammary glands along with formation and storage of triglycerides in adipose tissue and liver. Fatty acid synthesis is increased by activation and increased phosphorylation of acetyl-CoA carboxylase, while fat oxidation is suppressed by inhibition of carnitine acyltransferase. Triglyceride synthesis is stimulated by esterification of glycerol phosphate, while triglyceride breakdown is suppressed by dephosphorylation of hormone sensitive lipase. Cholesterol synthesis is increased by activation and dephosphorylation of HMG Co-A reductase while cholesterol ester breakdown appears to be inhibited by dephosphorylation of cholesterol esterase. Phospholipid metabolism is also influenced by insulin.[69]

C. Protein Synthesis

Insulin promotes protein synthesis in a range of tissues. There are effects on transcription of specific mRNA, as well as translation of mRNA into proteins in the ribosomes. Examples of enhanced mRNA transcription include the mRNA for glucokinase, PK, fatty acid synthase and albumin in the liver, pyruvate carboxylase in the adipose tissue, casein in the mammary gland and amylase in the pancreas. Insulin action decreases mRNA for liver enzymes such as carbamoyl phosphate synthetase, a key enzyme in the urea cycle. Effects on translation are widespread and infl uenced by both insulin itself and by various growth factors, e.g. IGF-1.[64, 69]

2.9 Diagnostic criteria for diabetes mellitus

2.9.1 WHO (2002) Criteria for the Diagnosis of Diabetes Mellitus (*ICMR guidelines also have the same diagnostic criteria for India*)[73]

S.No	Categories of Hyperglycemia	Glucose Concentrations mmol/l (mg/dl) Plasma
1.	Diabetes Mellitus	
	Fasting	≥7.0 (≥126)
	2-hour post glucose load (75g)	≥11.1 (≥200)
2.	Impaired Glucose Tolerance (IGT)	
	Fasting	<7.0 (<126)
	2-hour post glucose load (75g)	≥7.8 and <11.1 (≥140 and <200)
3.	Impaired Fasting Glycemia (IFG)	
	Fasting	≥5.6 and <7.0 (≥100 and <126) (≥110 and <126) as per ICMR Guidelines 2005
	2-hour post glucose load (75g)	<7.8 (<140)

2.9.2 Diagnostic indices of diabetes mellitus

A.Glucose determination

The simplest indicator of the adequacy of control of carbohydrate metabolism of a patient is the blood glucose concentration. However glucose is rapidly metabolized in the body. Therefore, the glucose concentration reflects the immediate status of carbohydrate metabolism and does not allow a retrospective or prospective evaluation of glucose metabolism.

Glucose is measured in different specimens, including

- whole blood (capillary or venous blood)
- haemolysate
- plasma
- serum
- de-proteinized blood

- urine
- CSF

Blood Glucose

The pathological entity of blood glucose is the plasma glucose concentration that is the glucose to which organ systems are exposed. Some glucose measurements detect plasma glucose directly (by electrode) and do not rely on a precise volume of plasma being applied. Plasma can also be prepared from whole blood by centrifugation but erythrocytes will continue to metabolize glucose thus lowering the concentration measurable unless glycolysis is inhibited. This is mainly achieved by use of fluoride bulb or vacutainer for collection of blood samples.

Whole blood glucose concentration is also affected by the concentration of protein (mainly haemoglobin - 8-18 percent) in the sample. For this reason whole blood concentrations are 12 to 15 percent lower than plasma concentrations by a variable amount, and plasma glucose is the preferred measure. Finger prick blood samples used for immediate testing on reagent strips or electrode sensors depend on the concentration of glucose in the plasma fraction, but such systems may be calibrated by the manufacturer to plasma or whole blood standards.

The plasma glucose concentration of importance at peripheral organ systems is the arterial concentration and this (or rather arterialized glucose) is the measure of preference in some research studies. Capillary blood glucose concentrations will be a good approximation to this provided tissue perfusion is good. Venous blood will have lower glucose concentrations than arterial blood (and thus capillary blood) but the effect is not large except where glucose disposal from the blood is high (after a meal due to insulin, or during exercise) and the sample is taken proximal to a muscle bed (eg from the ante-cubital fossa).

In post-prandial state or during glucose load (OGTT) capillary blood glucose levels are approximately 1.0 mmol/l (approximately 0.20 g/L) higher than in whole venous blood.

In whole blood glycolysis decreases the glucose concentration by 5–7 percent per hour at room temperature. Serum glucose once separated from erythrocytes remains stable at room temperature up to 8 hrs, or for up to 72 hrs at 4°C.

When collecting and transporting blood for glucose analysis it is important to inhibit enzymatic degradation of blood glucose. Glycolysis in whole blood is inhibited by sodium fluoride (6 g/L blood) or maleinimide (0.1 g/L blood). As anticoagulant EDTA (1.2-2 g/L blood) is used.

Cerebrospinal fluid (CSF) should be analyzed for glucose as soon as possible.

Blood glucose collection and stability

Collection:
1. Capillary blood
2. Venous blood
3. Plasma
4. De-proteinized blood
5. Haemolysate (digitonin, maleinimide)

Stability of specimen:

Venous blood: at 20°C: decrease of 10-15 percent/h at 4°C: decrease of 20 percent in 24 h
Stabilizer: NaF (6 g/L) + Maleinimide (0.1 g/L blood), EDTA (1,2-2 g/L) or EDTA + maleinimide
Plasma/serum: at 20 °C: decrease of 15 percent in 24 h
Deproteinized serum: stable over days and weeks

Interferences: Anticoagulants, drugs, glutathione, ascorbic acid, α-methyldopa

Pre-analytical effects: Posture, exercise, food ingestion, smoking, transport/preservation of specimen

Methods for blood glucose determination

Several methods are available for glucose determination. The methods for glucose analysis are the following:

Chemical methods	**Enzymatic methods**
Ortho-toluidine	Hexokinase-G6PDH
Neocuproine	Glucose dehydrogenase
Ferricyanide	Glucose oxidase-peroxidase (ABTS)
	Glucose oxidase (GOD) with other indicator reactions

For the chemical oxidation/reduction methods (neocuproine method, ferricyanide method) and the o-toluidine method the reagent costs are low. Although these methods are less specific, they are still useful and valid. The enzymatic analysis of glucose is more specific. However the enzymatic methods are also more expensive. The enzymatic reference method for glucose is the hexokinase/G6PDH method. The glucose dehydrogenase method has comparable analytical performance. The glucose oxidase methods performing slightly less well, since reducing substances may interfere with the peroxidase step. Nevertheless the GOD methods are most frequently used for convenience and economic reasons.[73]

The reference intervals of the three enzymatic methods for glucose in blood of fasting adults are:

Serum/plasma, Whole blood	
Hexokinase/G6PDH:	4.4 – 5.5 mmol/L (80–100 gm/l)
Glucokinase:	4.4 – 5.5 mmol/L (80–100 gm/l)
GOD/POD:	5.0 – 6.1 mmol/L (90–110 gm/l)

The concentration of glucose in cerebrospinal fluid is about 60 percent of the plasma value. If CSF is contaminated with bacteria or additional cells, the glucose concentration may be much lower.

B.Glycated proteins

Proteins react spontaneously in blood with glucose to form glycated derivatives. This reaction occurs slowly under physiological conditions and without the involvement of enzymes. The extent of glycation of proteins is controlled by the concentration of glucose in blood and by the number of reactive amino groups present in the protein that are accessible to glucose for reaction. All proteins with reactive sites can be glycated and the concentration of the glycated proteins that can be measured in blood is a marker for the fluctuation of blood glucose concentrations during a certain period. From a clinical diagnostic point glycated proteins with a longer life time in blood are of interest, since they reflect the exposure of these proteins to glucose for longer periods.

Glycated haemoglobin

The life span of haemoglobin in vivo is 90 to120 days. During this time glycated haemoglobin forms, being the ketoamine compound formed by combination of haemoglobin A and glucose. Several sub fractions of glycated haemoglobins have been isolated. Of these, glycated haemoglobin, a fraction HbA1c is of most interest serving as a retrospective indicator of the average glucose concentration over the previous 8 to 10 weeks.[73]

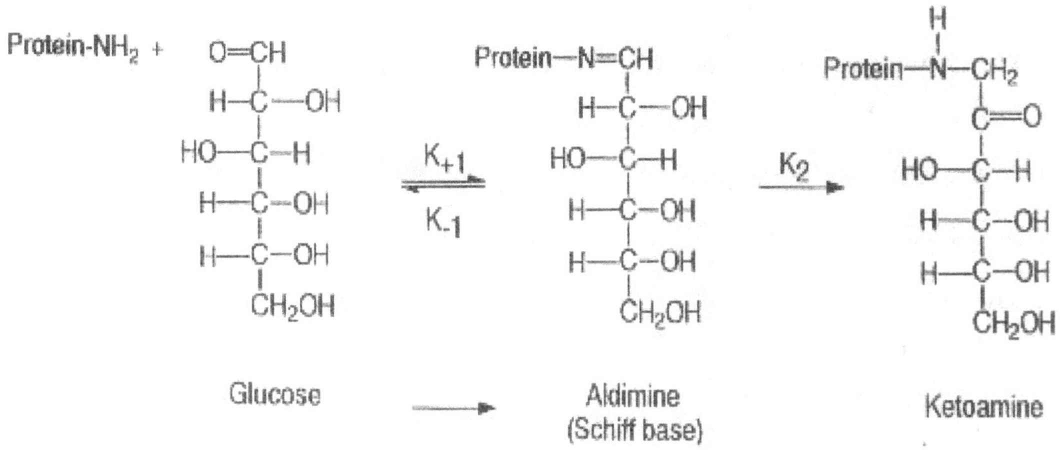

Table 2.2 Different methods used for estimation of HbA1c[73]

S.No.	Procedure	Principle	Analyte	Sample (blood)	Comment
1.	Column chromatography (macro-column)	Ion exchange chromatography	HbA1a, HbA1b, HbA1c	100 µl	The aldimine form is partially determined, interference by HbF, HbS, HbC, and acetaldehyde adducts
2.	Micro-column	Ion exchange chromatography	HbA1	100 µl	Interference by Hb variants, temperature and pH sensitive
3.	HPLC	Ion exchange chromatography	HbA1a, HbA1b, HbA1c	10 – 400 µl	Interference by Hb variants, temperature and pH sensitive
4.	FPLC	Ion exchange chromatography	HbA1a, HbA1b, HbA1c, aldimine	20-100 µl	Best separation of HbA1c and the aldimine;
5.	Thiobarbituric acid	Hydrolytic cleavage and colourimetric determination of ketohexoses	HbA1c	2-4 ml	Only detects the ketoamine form, the aldimine is eliminated; also reacts with sialic acid
6.	Electrophoresis	Electro-endosmosis	HbA1c	20 µl	
7.	Isoelectric focussing	pH gradient 5 – 6.5	HbA1c, aldimine	10 µl	Detection of abnormal haemoglobins
8.	Affinity chromatography	Phenylboronate column			
9.	Immuno-chemical methods	Specific antibodies (monoclonal, polyclonal) in EIA, immunotur-bidimetry	HbA1c, HbA2c	10-50µl	No interference by Hb variants; glycated HbA2, HbS1c, are detected but not HbF1c

Table 2.3 Reference intervals of HbA1c[73]

S.No	Method	Brand name	Indicator	Reference interval (percent)
1.	Affinity chromatography	GHb Imx	tHb HbA1c	tHb 4.8 – 7.8 HbA1c 4.4 – 6.4
2.	Affinity chromatography	Glyc-Affin	tHb	tHb 4.0 – 8.0
3.	Affinity chromatography	BM HbA1	HbA1	HbA1 5.0 – 8.0
4.	Affinity chromatography	Glyc-Hb	tHb	tHb 5.0 – 8.0
5.	Affinity chromatography(minicolumns)	HbA1 mini column test	HbA1	HbA1 3.4 – 6.1
6.	Agarose gel electrophoresis	DIATRAC	HbA1c	HbA1c 3.3 – 5.6
7.	Immunoturbidimetry polyclonal antibody	TinaQuant HbA1c	HbA1c	HbA1c 4.3 – 5.8 HbA1c 3.6 – 5.3
8.	EIA, monoclonal antibody	DAKO HbA1c	HbA1c	HbA1c 2.8 – 4.9 HbA1c 4.5 – 5.9
9.	Immunoturbidimetry, monoclonal antibody	DCA 2000	HbA1c	HbA1c 4.2 – 6.3
10.	Immunoturbidimetry, monoclonal antibody	Unimate	HbA1c	HbA1c 4.5 – 5.7
11.	Ion exchange chromatography (microcolumns)*	HbA1c microcolumn test	HbA1c	HbA1c 4.2 – 5.9
12.	HPLC ion exchange chromatography	DIAMAT	HbA1 HbA1c	HbA1 5.1 – 7.3 HbA1c 4.3 – 6.1
13.	HPLC ion exchange chromatography	HS-8	HbA1 HbA1c	HbA1 5.0 – 7.8 HbA1c 4.4 – 5.7
14.	HPLC ion exchange chromatography	L-9100	HbA1 HbA1c	HbA1 4.5 – 6.0 HbA1c 3.4 – 4.7

C. Estimation of Lipid Profile

Introduction

Lipids are a group of fats and fat-like substances that are important constituents of cells and sources of energy. A lipid profile measures the level of specific lipids in the blood.

A lipid profile typically includes estimations of:

- Total Cholesterol-This test measures all of the cholesterol in all the lipoprotein particles.
- High-density lipoprotein cholesterol (HDL-C) - Measures the cholesterol in HDL particles; often called "good cholesterol" because it removes excess cholesterol and carries it to the liver for removal.
- Low density lipoprotein cholesterol (LDL-C) - Calculates the cholesterol in LDL particles; often called "bad cholesterol" because it deposits excess cholesterol in walls of blood vessels, which can contribute to atherosclerosis. Usually, the amount of LDL cholesterol (LDL-C) is calculated using the results of total cholesterol, HDL-C and triglycerides.
- Triglycerides- Measures all the triglycerides in all the lipoprotein particles; most is in the very low-density lipoproteins (VLDL).

Some other information may be reported as part of the lipid profile. These parameters are calculated from the results of the tests identified above.

- Very low density lipoprotein cholesterol (VLDL-C) - Calculated from TG/5; this formula is based on the typical composition of VLDL particles.
- Non-HDL-C - Calculated from total cholesterol minus HDL-C.
- Cholesterol/HDL ratio - Calculated ratio of total cholesterol to HDL-C.

An extended profile (or advanced lipid testing) may also include low-density lipoprotein particle number/concentration (LDL-P). This test measures the number of LDL particles, rather than measuring the amount of LDL-cholesterol. It is thought that this value may more accurately reflect heart disease risk in certain people.

Different methods for estimation of lipid profile

1. Method for estimation of Total cholesterol

i) Chemical methods

1. p-toluenesulphonic acid and acetic anhydride method
2. Ferric chloride method
3. Liebermann-Burchard colorimetric method
4. Trinder's method

Reference interval: 50-250 mg/dl

ii) Enzymatic method

CHOD – POD Method, End point

Reference interval for Total Cholesterol
Desirable: Less than 200 mg/dl(5.18 mmol/l)
Borderline high: 200-239 mg/dl (5.18 to 6.18 mmol/l)
High: 240 mg/dL (6.22 mmol/l) or higher [78]

2. Methods for estimation of Triglycerides (TG method) [74]

i) Chemical methods

1. Isopropanol method

Reference interval: 25-160 mg/dl

ii) Enzymatic method

1. Glycerol kinase
2. Glycerol dehydrogen method

Reference interval:
Fasting Triglycerides
Desirable: Less than 150 mg/dl (1.70 mmol/l)
Borderline high: 150-199 mg/dl(1.7-2.2 mmol/l)
High: 200-499 mg/dl (2.3-5.6 mmol/l)
Very high: Greater than 500 mg/dl (5.6 mmol/l) [78]

3. Methods for estimation of HDL

HDL-Cholesterol (HDL-C) by Homogenous Enzymatic Direct Assay[75]

HDL Cholesterol	
Low level, increased risk:	Less than 40 mg/dL (1.0 mmol/l) for men Less than 50 mg/dl (1.3 mmol/l) for women
Average level, average risk:	40-50 mg/dL (1.0-1.3 mmol/l) for men Between 50-59 mg/dl (1.3-1.5 mmol/l) for women.
High level, less than average risk:	60 mg/dL (1.55 mmol/l) or higher for both men and women. [78]

4. Methods for estimation of LDL (LDL estimation) [76]

A. The most common approach to determining LDL-C in the clinical laboratory is the Friedewald calculation, which estimates LDL-C from measurements of TC, triglycerides (TGs) and HDL-C.

B. LDL-Cholesterol (LDL-C) by Homogenous Enzymatic Direct Assay[77]

Reference interval

LDL Cholesterol	
Optimal:	Less than 100 mg/dl (2.59 mmol/l)
Near/above optimal:	100-129 mg/dl (2.59-3.34 mmol/l)
Borderline high:	130-159 mg/dl (3.37-4.12 mmol/l)
High:	160-189 mg/dl (4.15-4.90 mmol/l)
Very high:	Greater than 190 mg/dl (4.90 mmol/l)[78]

5. Methods for estimation of Apo B

Immunoturbidometric methods

Reference interval

Adults: 50 - 155 mg/dl

D. Glucose tolerance test

Different glucose tolerance tests are used to diagnose type 2 diabetes or impaired glucose tolerance

1. Oral glucose tolerance test (OGTT)

OGTT is a challenge test to examine the efficiency of the body to metabolise (tolerance) glucose. It provides information on latent diabetes states. The OGTT distinguishes metabolically healthy individuals from people with impaired glucose tolerance and those with diabetes. The OGTT is more sensitive than FPG for the diagnosis of diabetes. Nevertheless the final diagnosis of diabetes should not be based on a single 2 hrs post-load glucose >11,1 mmol/l (>200 g/L) but should be confirmed in subsequent days (FPG and/or casual glucose estimation).

The OGTT is not used for the monitoring of day to day blood glucose control, which is done by HbA1c and repeated glucose measurement. It is not for routine diagnostic use but it is used in epidemiological population studies, but is not recommended or necessary for routine diagnostic use.

Preparation of the patient:

- Three days unrestricted, carbohydrate rich diet and activity.
- No medication on the day of the test.
- 12-h fast.
- No smoking.

Glucose load: Adults 75 g in 300 – 400 ml of water.

Children: 1.75 g/Kg up to 75 g glucose

Solutions containing glucose and oligosaccharides are commercially available.

Plasma glucose sampling:

- 10 min before glucose load
- Every 30 minutes after glucose load
- Last sample at 150 min after glucose load

- Urine glucose can be additionally measured in case of hyperglycaemia.

Evaluation:

Fasting plasma glucose 120 min glucose

- IFG 6.1-6.9 mmol/l (110-125 gm/l)
- IGT ≤7.0 mmol/l (<126 g/L) and 78-110 mmol/l (140-199 gm/l)
- Diabetes ≥7.0 mmol/l (>126 g/L) or >111 mmol/l (>200 gm/l)

These values are for the preferred measure of plasma glucose; different values apply to whole blood or capillary blood glucose.

Comments:

The OGTT is affected by metabolic stress from a number of clinical conditions and drug treatments, such as:

- Major surgery
- Myocardial infarction, stroke, infections, etc
- Malabsorption
- Drugs (steroids, thiazides, phenytoin, oestrogens, thyroxine)
- Stress, nausea
- Caffeine, smoking [73]

2. Intravenous Glucose tolerance

A bolus of glucose (300 mg/kg in a 30 percent solution) was given within 60 sec into the antecubital vein. Blood was sampled from the contralateral antecubital vein at 0, 2, 4, 6, 8, 10, 20, 30, 40, 50, 60 and 75 min for assessment of the plasma glucose. [78]

3. Mini Glucose tolerance test

This is a simple test monitor development of diabetes in pregnancy—known as gestational diabetes. In mini GTT two samples are taken as per sampling procedure given below.

Plasma glucose sampling:

- 10 min before glucose load
- Last sample at 120 min after glucose load

Urine glucose can be additionally measured in case of hyperglycaemia

E. Other indices

1. Peptide

C-peptide is cleaved from proinsulin, stored in secretory granules, and eventually released into the bloodstream in amounts equimolar with those of insulin. [79-81]C-peptide has an essential function in the synthesis of insulin in that it links the A and B chains in a manner that allows correct folding and interchain disulfide bond formation. When C-peptide is removed from proinsulin by proteolytic processing, the COOH-terminal part of insulin's B-chain becomes exposed and free to assume an appropriate conformation for effective interaction with the insulin receptor. [82]

Methods used for estimation

- Radioimmunoassay
- ELISA
- Chemiluminescence
- Fluorescence

Reference Interval

0.8-3.5 ng/mL

Interpretation

C-peptide has been widely used as a test of beta cell function in human subjects in a variety of conditions including Type I diabetes,[83] hypoglycemia,[84] and surreptitious insulin administration.[85,86] It has also been suggested that the C-peptide/insulin molar ratio reflects changes in hepatic insulin extraction in obesity, cirrhosis, diabetes, and following glucose administration.[87-93]

2. Proinsulin

Proinsulin (9000 mol wt. peptide) is synthesized on the ribosomes of pancreatic islet β cells and then transferred to Golgi apparatus, where during the process of maturation of secretory granules, proinsulin is cleaved by endo-peptidases PC3 (type 1) and PC2 (type 2) into insulin (6000 mol wt) and C-peptide (3000 mol wt.). Normal cleavage sites are Arg. 31-32 and Lys 64, Arg 65 at the A Chain.

During this processing, conversion intermediates are produced – type1 intermediates in which C peptide/ A chain has been cleaved and C peptide remains attached to the B chain (des Lys64 Arg65 proinsulin); and type 2 intermediate in which C peptide / B chain has been cleaved and C peptide remains attached to the A chain (des Arg31 Arg32 proinsulin).[94]

Interpretation

Proinsulin is used to evaluate β-cell function, because its elevated plasma levels are considered a sign of the worsening of the pancreatic activity.[95,96] In particular, the fasting proinsulin-to-insulin molar ratio is used as a marker of β-cell impairment.[97] However, this ratio is based on the concentration of insulin and proinsulin in the periphery; thus it cannot directly indicate a possible defect at the β-cell level, because it can be affected by several other factors of insulin and proinsulin kinetics that can vary significantly in different conditions.[98]To provide an index of β-cell function based on estimation of both secretion and kinetics of the peptides, mathematical modeling is necessary.[99-101]

Methods for estimation of Insulin

- ELISA
- Chemiluminescence

Reference Interval

Component	Reference Interval
Insulin, Fasting	3-19 μIU/mL
Proinsulin, Intact	0-17 years: Not established 18 years and older: Less than or equal to 28.9 pmol/L
Proinsulin, Intact/Insulin Ratio Calculation	**Proinsulin, Intact/Insulin Ratio as Percent:** 0-17 years: Not established 18 years and older: 0.8-21.7 percent

In diagnosis of diabetes, a combination of laboratory and clinical tests can be used to monitor blood glucose control, detect onset and progression of diabetic complications, and predict treatment response.

F. Laboratory tests for glycemic control

These tests are divided into three types

1. **Short term glycemic control**:

   Short term glycemic control includes determination of fasting plasma glucose (FPG) measurement and oral glucose tolerance test (OGTT). These tests give variable results due to stress, illness, diet and physical activity.[102, 103]

2. **Long term glycemic control**

   Standardized hemoglobin A1c (HbA1c) assays reliably estimate average glucose levels over a longer term (2-3 months), do not require fasting or glucose loading, have less variability due to stress and illness.They are are more specific for identifying individuals at increased risk for diabetes.[103]

3. Tight glycemic control

Glycosylated hemoglobin remains the gold standard marker for assessing long-term glycemic control. However, it does not provide "real-time" information about individual hyperglycemic or hypoglycemic excursions. To the contrary, Self-monitoring of blood glucose (SMBG) reveals the immediate hour-to-hour blood glucose, which in people without diabetes, varies only fifty percent throughout the day but may vary up to10-fold in patients with diabetes. Real-time information provided by SMBG represents an important adjunct to A1c, because it can track fasting and postprandial hyperglycemia, detect glycemic excursions and hypoglycemia and ultimately provides on-the-spot information about the instant effects of food choice, physical activity, and medication on glycemic control. In fact, SMBG can aid in diabetes control by doing the following: facilitating the development of an individualized blood glucose profile, which can then guide doctors in treatment planning for an individualized diabetic regimen; offering patients with diabetes the ability to make appropriate day-to-day treatment choices in diet and physical activity as well as in insulin or oral hypoglycemic agents; and improving patient recognition of hypoglycemia. If the principle of data-driven feedback governs the treatment adaptation, SMBG could ultimately show effectiveness in significantly lowering A1c.[104] In pursuit of achieving near euglycemia while avoiding hypoglycemia, the importance of SMBG is widely appreciated and recommended as a routine part of management in patients with type 1 or insulin-treated type 2 diabetes, as well as in diabetic pregnancy, for both women with pregestational type 1 and gestational diabetes.[105] In patients with type 1 diabetes in the Diabetes Control and Complications Trial, it was clearly shown that SMBG in the context of multifactorial interventions is linearly correlated with reductions in A1c.[106] Most authorities recommend subjects with type 1 diabetes using multiple insulin injections or insulin pump therapy to perform more than three capillary glucose determinations per day, but ideally four to six [105] In patients with type 2 diabetes managed with noninsulin therapies or medical nutrition therapy, despite the lack of clear evidence linking SMBG to improved glycemic control, the adoption of this practice is quite common and constantly increasing. National guidelines unanimously recommend SMBG in insulin treated type 2 diabetes; however, there is a lack of consensus on the value of SMBG in non–insulin-treated type 2 diabetes

(NIT-DM) patients. Considering the increasing incidence of type 2 diabetes and the tremendous economic burden of the disease both in direct and indirect costs, but also the not neglect able cost of the supplies for performing SMBG, such a gap of a straightforward recommendation regarding SMBG applicability and the optimal frequency of measurements is at least query worthy.

Self-monitoring of blood glucose (SMBG) is useful for tracking short-term treatment responses in insulin-treated patients, but its usefulness is less clear in non–insulin-treated patients.[102]

Different laboratory tests are available for monitoring blood glucose control over the short, long, and intermediate term to help evaluate the effectiveness of a management plan. [102]

Part II

Metabolic syndrome

2.10. What Is Pre-Diabetes?

The term pre-diabetes has had a checkered history. Alberti,[107] states that it was first used to denote abnormalities of pregnancy (e.g., high– birth weight babies, hydramnios) or a strong family history of type 2 diabetes. However, in 1980, the World Health Organization (WHO)[108] discarded the term largely because many subjects with borderline glucose levels do not convert to diabetes and because many would be alarmed unnecessarily. In 2005, the American Diabetes Association (ADA) reintroduced the term pre-diabetes to cover impaired glucose tolerance (IGT) and impaired fasting glucose (IFG) but not other risk factors for diabetes.[109] In 2008, WHO's diabetes task force again repudiated the term and discouraged its use.[110] Instead, they suggested "intermediate hyperglycemia" to signify IGT and IFG. The ADA nonetheless continues to use pre-diabetes and defines it as IFT, IGT and now includes subjects with HbA1c in the range of 5.7 percent to 6.4 percent.[111] Any definition of pre-diabetes that is restricted to IGT and/or IFG fails to include other risk factors for diabetes such as a family history of type 2 diabetes or the metabolic syndrome. Another telling criticism of the term pre-diabetes is that the many subjects with either IFG or IGT will not progress to type 2 diabetes. For this reason, another name was thought to be be preferable. The WHO task force's "intermediate hyperglycemia" so far has not been widely adopted.[110] Another possible name is "borderline diabetes," but this term is not currently recommended and has no formal definition.

Thus the term "prediabetes" places individuals at high risk of developing diabetes and its complications. Not all individuals with prediabetes will necessarily progress to diabetes. Indeed, a significant proportion of people who are diagnosed with IFG or IGT will revert to normoglycemia. People with prediabetes, particularly in the context of the metabolic syndrome, would benefit from cardiovascular risk factor modification. While people with prediabetes do not have the increased risk for microvascular disease as seen in diabetes, they are at risk for the development of diabetes and CVD.[112] IGT is more strongly associated with CVD outcomes than is IFG. Individual's identified as having both IFG and IGT are at higher risk for diabetes as well as CVD. While there is no worldwide consensus on the definition of IFG,[113,114] the Canadian Diabetes Association defines IFG as an FPG value of 6.1 to 6.9 mmol/l due to the higher risk of developing diabetes in these individuals compared to defining IFG as an FPG value of 5.6 to 6.9 mmol/L. [114] While there is a continuum of risk for diabetes in individuals with HbA1c levels between 5.5% and 6.4%, population studies demonstrate that HbA1c levels of 6.0%to 6.4% are associated with a higher risk for diabetes compared to levels between 5.5% and 6.0%.[115]While the American Diabetes Association defines prediabetes as an HbA1c between 5.7% and 6.4%, the Canadian Diabetes Association has based the definition on a higher risk group and includes an HbA1c of 6.0% to 6.4% as a diagnostic criterion for prediabetes.[103] However, HbA1c levels below 6.0% can indeed be associated with an increased risk for diabetes.[115] The combination of an FPG of 6.1 to 6.9 mmol/L and an HbA1c of 6.0% to 6.4% is predictive of 100% progression to type 2 diabetes over a 5-year period.[116]

2.10.1 Metabolic syndrome

Prediabetes and type 2 diabetes are often manifestations of a much broader underlying disorder,[117] including the metabolic syndrome a highly prevalent, multifaceted condition characterized by a constellation of abnormalities that include abdominal obesity, hypertension, dyslipidemia and elevated blood glucose. Individuals with the metabolic syndrome are at significant risk of developing CVD. While metabolic syndrome and type 2 diabetesoften coexist, those with metabolic syndrome without diabetes are at significant risk of developing diabetes in future. In 2009, a harmonized definition of the metabolic syndrome was established, with at least 3 or more criteria required for diagnosis NCEP ATP III.[118]

The insulin resistance condition comprises a spectrum of disorders, with hyperglycemia representing one of the most readily diagnosed features. The metabolic syndrome, the insulin resistance syndrome, or syndrome X are terms used to describe a constellation of metabolic derangements that includes insulin resistance, hypertension, dyslipidemia (decreased HDL and elevated triglycerides), central or visceral obesity, type 2 diabetes or IGT/IFG, and accelerated cardiovascular disease.[119]

Although the term MS has become widely used since its inception in 2001 by the NCEP: ATP III [9] the concept of "clustering" metabolic disorders and CVD risk factors has been discussed in the scientific literature for many decades. In fact, recent reviews have noted that independent scientists published reports of the association between diabetes mellitus and hypertension as early as the 1920s, [120] when Kylin, [121] documented a connection between hypertension, hyperglycemia and gout. While the primer for understanding visceral adiposity did not occur until nearly 30 yrs later,[122] by the early 1990s visceral obesity was fully appreciated as a component of the insulin resistance syndrome.[123] In 1980, the seminal work of Margaret Albrink,[124] focused on the relationship between obesity, hypertriglyceridemia and hypertension.[125] It was not until 1988 when Reaven,[108] in his landmark Banting Lecture, coined the term "Syndrome X" to describe the proposed interrelationships between resistance to insulin-stimulated glucose uptake, hypertension, type 2 diabetes and CVD. During the ensuing 10 yr, Syndrome X and other terms were used to describe the clustering of cardiovascular and metabolic risk factors, including "deadly quartet",[126] and the "insulin resistance syndrome".[126-129]

2.10.2 Pathophysiology of Metabolic Syndrome

i) Insulin resistance: a conceptual prologue

Insulin is a pleiotropic molecule that has effects on amino acid uptake, protein synthesis, proteolysis, adipose tissue triglyceride lipolysis, lipoprotein lipase activity, very low density lipoprotein (VLDL) triglyceride secretion, muscle and adipose tissue glucose uptake, muscle and liver glycogen synthesis and endogenous glucose production. Individuals are generally defined as insulin sensitive or insulin resistant by their response to an oral or IV glucose or insulin stimulus.[130] Characteristics of the insulin-sensitive phenotype include a normal body weight,[131] without abdominal or visceral obesity [123,132] being moderately active,[133] and consuming a diet low in saturated fats. Alternatively, insulin-resistant individuals demonstrate impaired glucose metabolism or tolerance by an abnormal response to a glucose challenge, elevated fasting glucose levels and/or overt hyperglycemia, or reductions in insulin action after IV administration of insulin (euglycemic clamp technique) with decreased insulin-mediated glucose clearance and/or reductions in the suppression of endogenous glucose production. In general, the characteristics of this phenotype are more likely to include being overweight or obese,[131,134] being sedentary [133] and consuming a diet high in total or saturated fats.[135] Insulin sensitivity, however, is not a simple dichotomy of being insulin sensitive or insulin resistant, but rather exists on a continuum. Moreover, the ability of the pancreas to secrete insulin in response to a glucose challenge may also reflect insulin resistance at the level of the β-cell. To define this, Bergman, [134] proposed the disposition index, a quantitative measure that describes the relationship between β-cell sensitivity and insulin sensitivity.[136] In metabolically normal individuals, changes in insulin sensitivity are accompanied by compensatory alterations in the response of the β -cell to glucose. In practice, disposition index is defined as the product of the insulin sensitivity index and β-cell function as measured by the acute insulin response to glucose, a relationship that is typically plotted as an inverse hyperbola. The movement along the continuum is more complicated than the model implies and the molecular mechanism(s) by which insulin sensitivity and β -cell function are coregulated to create a homeostatic environment are not well understood.

ii) Obesity as a "driving force" in the prevalence of insulin Resistance

The worldwide increase in the prevalence of obesity in the recent decades is startling and is likely a cause of the rising incidence of insulin resistance and the metabolic syndrome [108, 125, 137-139] as well as CVD and type 2 diabetes.[125] Although not all overweight or obese individuals are metabolically unhealthy, the majority are insulin resistant.[140] Indeed, many experts assert that the MS would never have been put forth if the obesity epidemic had not become the public health concern that it is today.[137] In particular, the combination of obesity, physical inactivity and consumption of an atherogenic diet is believed to lead to insulin resistance. [141] In this state of insulin resistance, normoglycemia is initially maintained by a modest increase in β-cell mass and/or an increase in insulin secretory capacity. [108,142] Although the mechanism for this compensation is unclear, there is recent evidence supporting glucose signaling as a dominant force in this process; [143] it is also acknowledged that genetic factors may be involved.[144,138] However, the loss of insulin secretory capacity in the natural history of type 2 diabetes is likely an admixture of β-cell dysfunction in addition to reductions in β -cell mass.[146] If the increasing β -cell function and/or mass is a successful long term compensatory mechanism to obesity and insulin resistance, T2DM could be prevented for an undetermined amount of time, despite hyperinsulinemia as a consequence.

iii) Insulin resistance in adipose tissue

Adipose tissue insulin resistance appears to be important to the pathophysiology of the MS. [108,125, 137,147] Specifically, a larger, expanded adipose tissue mass often results in an increased turnover of free fatty acids (FFAs).[148,149] In the setting of insulin resistance and expanded adipose tissue triglyceride stores, the process of FFA mobilization (lipolysis) from stored adipose tissue triglyceride is accelerated.[125,150] Under normal conditions, insulin inhibits adipose tissue lipolysis; however, in the setting of insulin resistance, insulin is unable to properly suppress lipolysis, resulting in relatively more FFA being liberated into the plasma.[137] Although it is well accepted that this process is mediated by hormone-sensitive lipase (HSL),[151] recent evidence points to adipose triglyceride lipase as playing an additional role and collectively these two hormones account for 95% of triglyceride hydrolysis.[152] In obese subjects, insulin resistance and hyperinsulinemia are strongly associated with decreased adipose triglyceride lipase and HSL mRNA and protein expression, an effect found to be independent of fat mass.[153] There is also evidence supporting a genetic predisposition for insulin resistance and type 2 diabetes linked to the HSL gene.[154] Not only does insulin resistance appear to cause FFA to rise but elevated FFA levels also appear to cause insulin resistance. Substantial evidence has accumulated to suggest that the visceral depot contributes to increased FFA turnover and insulin resistance. [155-158] Specifically, visceral adipocytes are more sensitive to catecholamine-stimulated lipolysis than sub cutaneous adipocytes.[159] Because the venous drainage of the visceral adipose tissue depot is directly into the portal system,[158] it has been hypothesized that in visceral obesity the liver is bathed with fatty acids and consequently becomes insulin resistant ("portal theory").[137,160]

iv) Insulin resistance in the liver

The liver plays a major role in substrate metabolism. Increases in FFA flux have been shown in numerous models to impair hepatic insulin action.[160] This includes increases in hepatic glucose output, the synthesis of proinflammatory cytokines and major changes in lipoprotein metabolism. In the liver, the increased FFA flux must be oxidized or stored.

Insulin, under normal physiological conditions, increases the gene expression of a number of enzymes central to triglyceride biosynthesis,[161] but also reduces VLDL triglyceride and apolipoprotein (apo) B production and secretion, an effect largely attributable to reductions in adipose tissue lipolysis.[162] Another intrahepatic effect of insulin is to enhance Apo B degradation.[163] In the liver of insulin resistant patients, FFA flux is high, triglyceride synthesis and VLDL is also more.[164] For the most part, it is believed that the dyslipidemia associated with insulin resistance is a direct consequence of increased VLDL secretion by the liver.[163] This may explain why increases in plasma Apo B are variably associated with the metabolic syndrome in the absence of increases in LDL-C.[165] In addition to overproduction of VLDL by the liver, alterations in lipoprotein lipase (LPL) have been associated with the MS. Hypertriglyceridemia is typically associated with reductions in HDL-C. This in part relates to the transfer of cholesteryl ester from the core of triglyceride-rich lipoproteins to HDL-C, a process catalyzed by cholesteryl ester transfer protein (CETP).[125,137,166] This generates a smaller, triglyceride-rich HDL-C that is a better substrate for hepatic lipase, which results in a particle that is more rapidly cleared by the kidney.[167] Of interest, the MS has been associated with increased CETP mass in men and possibly reduced LDL-C particle diameter in addition to reduced HDL-C.[168]

v) Insulin resistance in muscle

In muscle, increased plasma FFA disrupt the glucose-fatty acid cycle.[148,169] The predominant defect in insulin action in skeletal muscle relates to an inhibitory effect of this increase in plasma FFA on insulin-mediated glucose transport.[170-172] It has also been hypothesized that triglyceride accumulation in skeletal muscle plays a direct role in the etiology of insulin resistance. There is also evidence that the degree of whole body insulin sensitivity is inversely correlated with intra muscular triglyceride content.[147, 173]

vi) Hypertension and insulin resistance

The relationship between insulin resistance and hypertension has been established [174-176] and relates to several potentially different mechanisms. First, it is important to note that insulin is a vasodilator when given intra vascular to people of normal weight,[177] with secondary effects on sodium reabsorption in the kidney.[178] In the setting of insulin resistance, the vasodilatory effect of insulin can be lost,[179] but the renal effect on sodium reabsorption preserved.[180] Fatty acids themselves can mediate relative vasoconstriction.[181] Moreover, the infusion of fatty acids into the portal vein activates the sympathetic nervous system and elevates blood pressure in rodents.[182] Insulin also increases the activity of the sympathetic nervous system,[183] an effect that might also be preserved in the setting of insulin resistance.[184] However, when assessed by concentrations of fasting insulin or the homeostatic model assessment (HOMA),[185] insulin resistance contributes only modestly to the increased prevalence of hypertension in the MS.[186] Because adipose tissue is a source of angiotensinogen,[187] it is not surprising to note the association of hyperaldosteronism with hypertension and the MS.[188]

Thus, metabolic syndrome in fact can be considered a pre-diabetic state. Compared with persons without metabolic syndrome, those with the syndrome have an approximate 5-fold increase in diabetes risk.[189] Although it might be assumed that this greater risk is due to IFG or IGT in patients with metabolic syndrome, Lorenzo et al [190] found that metabolic syndrome without pre-diabetes carries an approximate 5-fold increase in diabetes risk. For patients with IFG or IGT, the diabetes risk is 5 to 7 fold higher for than normoglycemia.[190] However, when pre-diabetes combines with metabolic syndrome; the risk is increased even more. We might ask why metabolic syndrome alone is a strong predictor of diabetes. For one reason, similar percentages of the population with metabolic syndrome and pre-diabetes are obese; [191] thus; a metabolic syndrome like pre-diabetes is an insulin-resistant state, which imparts diabetes risk. The metabolic syndrome may further predispose to beta cell dysfunction through lipotoxicity.[192] Thus; it should not be surprising that the metabolic syndrome associates with a high risk for diabetes.

vii) Should pre-diabetes be called a disease?

All the dialectologists have an general agreement that diabetes deserves to be called a disease and not just a risk factor for CVD. This agreement rests on the broad range of complications accompanying hyperglycemia: macrovascular diseases, microvascular diseases and various forms of neuropathy, among others. Patients with diabetes are prone to coronary heart disease, stroke, and peripheral vascular disease along with retinopathy, chronic kidney disease, bladder dysfunction, erectile dysfunction, orthostatic hypotension, gastroparesis and skin disorders. In a variety of ways, these conditions can be attributed to prolonged hyperglycemia. Although in the sense that diabetes is a "risk factor" for these disorders, their aggregation justifies calling diabetes a disease (refer figure 2.11). To call pre-diabetes per se a disease is less clear-cut. All of the preceding complications are less common, although they have been reported in some patients.[193]

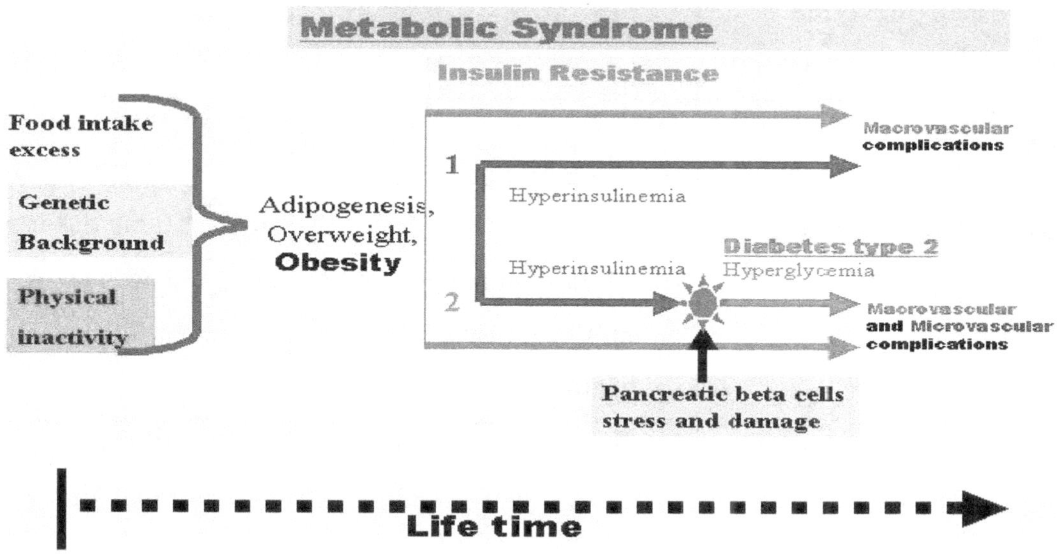

Figure 2.11. The relationship between metabolic syndrome, insulin resistance, hyperinsulinemia and hyperglycemia (overt type 2 diabetes). An insulin-resistant state following nuclear peroxisome proliferator activated receptors (PPAR) deactivation is the key phase of metabolic syndrome initiation. Afterwards, there are 2 principal pathways of metabolic syndrome development: 1) With preserved pancreatic beta cells function and insulin hypersecretion which can compensate for insulin resistance. This pathway leads mainly to the macrovascular complications of metabolic syndrome; 2) with massive damage of pancreatic beta cells leading to progressively decrease of insulin secretion and to hyperglycemia (e.g. overt type 2 diabetes). This pathway leads both to microvascular and macrovascular complications. Time-related scheme.[194]

2.10.3 Diagnostic criteria for Metabolic Syndrome

Table 2.4 Criteria for the Metabolic Syndrome definitions

World Health Organisation (WHO) criteria 1998 [28]	**European Group for the Study of Insulin Resistance EGIR, 1999**[195]	**National Cholesterol Education Program's Adult Treatment Panel III NCEP:ATPIII, 2001**[9]
High insulin levels, IFG or IGT and two of the following:	Top 25 percent of the fasting insulin values among non diabetic individuals and two of the following:	Three or more of the following:
Abdominal obesity: WHR >0.9 BMI ≥30 kg/m^2 WC > 37 inches	WC: 94 ≥ cm for men, 80 cm ≥ for women	WC: >40 inches for men, >35 inches for women
Lipid panel with TGs > 150 mg/dl HDL-C <35 mg/dl	Triglycerides ≥2.0 mmol/l HDL-C ≥1.0 mg/dl	Triglycerides ≥150 mg/dl HDL-C: <40 mg/dl for men, <50 mg/dl for women
BP >140/90 mm Hg	BP ≥140/90 mm Hg or antihypertensive medication	BP ≥130/85 mm Hg
	Fasting glucose ≥6.1 mmol/l	FPG ≥110 mg/dl

American Association of Clinical Endocrinologists, AACE, 2003 [196]	International Diabetes Federation IDF, 2006[197]
IGT and two or more of the following:	Aperson to be defined as having the metabolic syndrome they must have: Central obesity (defined as waist circumference ≥ 94cm for Europid men and ≥ 80cm for Europid women, with ethnicity specific values for other groups) plus any two of the following four factors:
Triglycerides ≥150 mg/dl HDL-C < 40 mg/dl for men, <50 mg/dl for women	TG level ≥ 150 mg/dl (1.7 mmol/L) HDL-C < 40 mg/dl in males < 50 mg/dl in females
BP ≥130/85 mm Hg	BP ≥130/85 mm Hg
	FPG ≥ 100 mg/dl

2.10.4 Dignostic methods for assessing insulin resistance and β-cell function

i) Hyperinsulinemic euglycemic glucose clamp

The hyperinsulinemic euglycemic glucose clamp technique has been described as the gold standard method for quantifying insulin sensitivity.[198] It is the reference method for quantifying insulin sensitivity in humans because it directly measures the effects of insulin in promoting glucose utilization under steady-state conditions in vivo.[198] Direct estimation of IR by means of the euglycemic clamp technique and insulin suppression test (IST) is experimentally demanding, complicated and impractical when large scale epidemiological studies are involved. These methods are laborious, painful and expensive, are therefore rarely used in larger-scale clinical research and as such are irrelevant for clinical practice. Consequently, over the years, a number of surrogate indices for insulin sensitivity or insulin resistance have been developed. The glucose clamp is difficult to apply in large scale investigations because of the chaotic procedure, which involves intra-venous infusion of insulin, taking frequent blood samples over a 3 h period and the continuous adjustment of a glucose infusion.[199]

ii) Homeostasis Model of Assessment (HOMA)

HOMA was first developed in 1985 by Matthews et al [200] It is a method used to quantify insulin resistance and beta cell function from basal (fasting) glucose and insulin (or C-peptide) concentrations. HOMA is a model of the relationship of glucose and insulin dynamics that predicts fasting steady-state glucose and insulin concentrations for a wide range of possible combinations of insulin resistance and β-cell function. Insulin levels depend on the pancreatic β-cell effect to glucose concentrations while, glucose concentrations are regulated by insulin-mediated glucose production via the liver. Thus, deficient β-cell function will echo a diminished response of β-cell to glucose-stimulated insulin secretion.[185,200,201] Similarly, insulin resistance is reflected by the diminished suppressive effect of insulin on hepatic glucose production. The HOMA model has proved to be a robust clinical and epidemiological tool for the assessment of insulin resistance. HOMA describes this glucose-insulin homeostasis by means of a set of simple, mathematically-derived nonlinear equations. The approximating equation for insulin resistance has been simplified and uses a fasting blood sample. It is derived from the use of the insulin glucose product, divided by a constant. The product of FPG × FPI is an index of hepatic insulin resistance.[199]

HOMA-IR = (glucose × insulin)/22.5: Insulin concentration is reported in μU/L and glucose in mmol/l product of normal fasting plasma insulin of 5 μU/ml and the normal fasting plasma glucose of 4.5 mmol/L typical of a "normal" healthy individual = 22.5. Whereas the β-cell function is also calculated by another equation using fasting insulin and glucose values.[199]

HOMA1 – percent B = (20 × FPI)/(FPG - 3.5): On the other hand, HOMA β-cell is another calculated variable indicating the insulin activity. It is a marker of basal insulin secretion of pancreatic β-cells.[199]

HOMA β cell = 20 × fasting plasma insulin (μU/ mL)/FPG (mmol)-3:

Estimation with the help of HOMA model parallels equally with that of the euglycemic clamp method (r = 0.88).[201] HOMA-IR has been observed to have a linear correlation with the glucose clamp and minimal model estimates of insulin sensitivity/resistance in various studies of distinct populations. [201,203] Derived from a mathematical assessment of the interaction between β -cell function and IR, the HOMA model is used to compute steady-state insulin and glucose concentrations. C-peptide, a measure of insulin secretion (not insulin action), can be used in HOMA modelling of both β-cell function and IR.[199]

iii) QUICKI

QUICKI is an empirically-derived mathematical transformation of fasting blood glucose and plasma insulin concentrations that provides a consistent and precise index of insulin sensitivity with better positive predictive power.[198, 203-205] It is simply a variation of HOMA equations, as it transforms the data by taking both the logarithm and the reciprocal of the glucose-insulin product, thus slightly skewing the distribution of fasting insulin values. QUICKI has been seen to have a significantly better linear correlation with glucose clamp determinations of insulin sensitivity than minimal-model estimates; especially in obese and diabetic subjects.[203] It employs the use of fasting values of insulin and glucose as in HOMA calculations. QUICKI[204] is virtually identical to the simple equation form of the HOMA model in all aspects, except that a log transform of the insulin glucose product is employed to calculate QUICKI.

QUICKI = 1/[log (Insulin μU/mL) + log (Glucose mg/dL).[204]

QUICKI should not be considered, as a new model rather simply logs HOMA-IR, which explains the near perfect correlation with HOMA. It has similar drawbacks to the use of the HOMA equations, compared with the computer model. Given the similarities between QUICKI and HOMA, the two methods compare well. In conditions like diabetes, glucose intolerance and hyperlipidemia associated with insulin resistance, or with various combinations of these metabolic disorders. QUICKI index values have been observed to be lower when compared to those of healthy volunteers. Adult patients with a

QUICKI index below 0.357 (which is at the lower limit of 95% confidence limits in healthy people) tend to have a higher risk or frequently present with typical manifestations of metabolic syndrome.[205] Each laboratory should establish its own normal QUICKI range, since variations in insulin determinations of different laboratories is unavoidable.[199]

iv) Surrogate Markers

Oral glucose tolerance test

The oral glucose tolerance test (OGTT) is an easy test and is commonly used in medical practice to detect glucose intolerance as well as type 2 diabetes.[206] It involves the administration of glucose to find out how rapidly it is cleared from the blood stream. It implicates the efficiency of the body to utilize glucose after glucose load. During OGTT, after 8 to 10 hrs of fasting, blood glucose levels are determined at 0, 30, 60, and 120 min following a standard oral glucose load (75 gm).[206,207] It imitates the normal physiology of the glucose and insulin flux more closely than conditions of the other methods such as the glucose clamp, IST or the Frequently Sampled Intravenous Glucose Tolerance Test (FSIVGTT). Since glucose tolerance and insulin sensitivity are dissimilar conceptually, OGTT provides useful information about glucose tolerance but not insulin resistance. However, OGTT is also used to estimate other surrogate indices of insulin resistance.[199]

Impaired glucose tolerance offers few aberrations during OGTT. Firstly, rapid and continuous rise in plasma glucose concentration and secondly, lack of decline below 140 mg/dL in plasma glucose at 2 hrs after attaining peak value. Subjects with impaired fasting glucose (IFG) have higher FPG than individuals with normal glucose tolerance or impaired glucose tolerance (IGT).[208]

Fasting insulin measurement

Measurement of the fasting insulin level has long been considered the most practical approach for the measurement of insulin resistance. It correlates well with insulin resistance. A considerable correlation has been found between fasting insulin levels and insulin action as measured by the clamp technique. A substantial overlap between insulin-resistant and normal subjects is a constraint, as there is a lack of standardization of the insulin assay procedure. Nevertheless, with a reliable insulin assay, insulin resistance can be detected early, before clinical disease appears.[209] As glucose levels change rapidly in the postprandial state, the use of fasting insulin for estimating IR should be done after an overnight fast, since the variable levels of glucose confound the simultaneous measure of insulin. In healthy subjects, increased fasting insulin levels (with normal fasting glucose levels) correspond to insulin resistance. In this population 1/ (fasting insulin) can substituted for insulin sensitivity that decreases as subjects become more insulin resistant (and fasting insulin levels rise). However, it does not cover the inappropriately low insulin secretion in the face of hyperglycemia seen in diabetic subjects or glucose-intolerant subjects. Use of fasting insulin levels for assessment of IR is limited because of a high proportion of false-positive results and by lack of standardization. To overcome this issue, standardization of insulin assay has been recommended by the ADA Task Force, to be certified by a central laboratory.[208]A high plasma insulin value in individuals with normal glucose tolerance reflects insulin resistance and high insulin levels presage the development of diabetes.[209]

Part –III

Diabetic Complications and Endothelial dysfunction

2.11 Diabetic Complications and their Pathogenesis

The chronic complications of diabetes mellitus affect many organ systems and are responsible for the majority of morbidity and mortality. Chronic complications can be divided into vascular and nonvascular complications. The vascular complications are further subdivided into microvascular (retinopathy, neuropathy, and nephropathy) and macrovascular complications (coronary artery disease, peripheral vascular disease and cerebrovascular disease). Nonvascular complications include problems such as gastroporesis, sexual dysfunction and skin changes. As a consequence of its chronic complications, DM is the most common cause of adult blindness, a variety of debilitating neuropathies, cardiac and cerebral disorders. Treating the complications of diabetes costs more than controlling the disease. Early in the course of diabetes, intracellular hyperglycemia causes abnormalities in blood flow and increased vascular permeability. This reflects decreased activity of vasodilators such as nitric oxide, increased activity of vasoconstrictors such as angiotensin II and endothelin-1 and elaboration of permeability factors such as vascular endothelial growth factor (VEGF). In diabetic arteries, endothelial dysfunction seems to involve both insulin resistance specific to the phosphotidylinositol–3-OH kinase pathway and hyperglycemia.[210]

2.11.1 Diabetic retinopathy

Diabetic retinopathy occurs in 3/4 of all persons having diabetes for more than 15 years and is the most common cause of blindness. There is appearance of retinal vascular lesions of increasing severity, culminating in the growth of new vessels. Diabetic retinopathy is classified into two stages: non proliferative and proliferative. The non-proliferative stage appears late in the first decade or early in the second decade of disease and is marked by retinal vascular microneurisms, blot hemorrhages and cotton-wool spots and includes loss of retinal pericytes, increased retinal vascular permeability, alterations in regional blood flow and abnormal retinal microvasculature, all of which lead to retinal ischemia. In proliferative retinopathy there is the appearance of neovascularization in response to retinal hypoxia. The newly formed vessels may appear at the optic nerve and/or macula and rupture easily, leading to vitreous hemorrhage, fibrosis, and ultimately retinal detachment.[211]

2.11.2 Diabetic Neuropathy

About half of all people with diabetes have some degree of neuropathy, which can be polyneuropathy, mono-neuropathy, and/or autonomic neuropathy. In polyneuropathy there is loss of peripheral sensation which, when coupled with impaired microvascular and macrovascular junction in the periphery, can contribute to non-healing ulcers, the leading cause of non-traumatic amputation. There is thickening of axons, decrease in microfilaments and capillary narrowing involving small myelinated or non-myelinated C-fibers. It can occur both from direct hyperglycemia-induced damage to the nerve parenchyma and from neuronal ischemia leading to abnormalities of microvessels, such as endothelial cell activation, pericyte degeneration, basement membrane thickening and monocyte adhesion. Mono-neuropathy is less common than polyneuropathy and includes dysfunction of isolated cranial or peripheral nerves. Autonomic neuropathy can involve multiple systems, including cardiovascular, gastrointestinal, genitourinary, sudomotor and metabolic systems.[212]

2.11.3 Diabetic Nephropathy

This is a major cause of end-stage renal disease. There are glomerular hemodynamic abnormalities resulting in glomerular hyper-filtration, leading to glomerular damage as evidenced by microalbuminuria. There is overt proteinuria, decreased glomerular filtration rate and end-stage renal failure. Dysfunction of the glomerular filtration apparatus is manifested by microalbuminuria and is attributed to changes in synthesis and catabolism of various glomerular basement membrane macromolecules such as collagen and proteoglycans, leading to an increase in glomerular basement thickening. Another possible mechanism to explain the increase in permeability of the glomerulus is the increase in renal vascular endothelial growth factor (VEGF) levels observed in preclinical models of diabetes, since VEGF is both an angiogenic and a permeability factor.[213]

2.11.4 Cardiovascular morbidity and mortality

In diabetes mellitus there is marked increase in several cardiovascular diseases, including peripheral vascular disease, congestive heart failure, coronary artery disease and myocardial infarction and a one to fivefold increase in sudden death. The absence of chest pain (silent ischemia) is common in individuals with diabetes and a thorough cardiac evaluation is indicated in individuals undergoing major surgical procedures. Despite proof that improved glycemic control reduces microvascular complications in diabetes mellitus, it is possible that macrovascular complications may be unaffected or even worsened by such therapies. An improvement in the lipid profiles of individuals in the intensive group (lower total and low-density lipoprotein cholesterol, lower triglycerides) suggested that intensive therapy may reduce the risk of cardiac vascular mortality. In addition to coronary artery disease, cerebrovascular disease is increased in individuals with diabetes mellitus (threefold increase in stroke). Individuals with DM have increased incidence of congestive heart failure (diabetic cardiomyopathy). The etiology of this abnormality is probably multifactorial and includes factors such as myocardiac ischemia from atherosclerosis, hypertension and myocardial cell dysfunction secondary to chronic hyperglycemia. Though DM itself does not increase levels of LDL, LDL particles found in T2DM are more atherogenic and are more easily glycated and susceptible to oxidation.[214]

2.11.5 Vascular complications in Diabetes mellitus

Diabetes and its associated complications have become a public health problem of considerable magnitude. Cardiovascular disease causes most of the excess morbidity and mortality in diabetes mellitus. Adults with diabetes are at a 2 to 4 fold increased risk of cardiovascular events relative to those without diabetes.[215] Cardiovascular disease accounts for up to 80 percent of premature excess mortality in diabetic patients.[216] Vascular complications can be caused by micro- and macro-angiopathy. Retinal and renal microangiopathy cause diabetic retinopathy and nephropathy and microangiopathy of the vasa nervorum is important in the pathogenesis of neuropathy. Macroangiopathy in diabetes consists mainly of an accelerated form of atherosclerosis and affects the coronary, carotid and peripheral arteries, thus increasing the risk of myocardial infarction, stroke and diabetic foot disease.[217-220] Dysfunction of the vascular endothelium is regarded as an important factor in the pathogenesis of micro- and macro-angiopathy.[221, 222]

2.12 Endothelial Function and Dysfunction

The endothelium is the biological active inner layer of the blood vessels, which serves as an important locus of control of vascular and thus organ functions.[223] The endothelium actively regulates vascular tone and permeability, the balance between coagulation and fibrinolysis, the composition of the sub endothelial matrix, the adhesion and extravasation of leucocytes and inflammatory activity in the vessel wall. It also affects the functions of other cell types, such as vascular smooth muscle cells, platelets, leucocytes, retinal pericytes, renal mesangial cells and large artery macrophages. On the other hand, these cells can affect endothelial cells. To carry out its above-mentioned functions, the endothelium produces components of the extracellular matrix such as collagen and a variety of regulatory mediators, including NO (nitric oxide), prostanoids, ET-1 (endothelin-1), Ang II (angiotensin II), t-PA (tissue-type plasminogen activator), PAI-1 (plasminogen activator inhibitor-1), vWF (von Willebrand factor), adhesion molecules and cytokines. The production of these moieties is responsive to various stimuli.[224] Normally, the endothelium actively decreases vascular tone, limits leucocyte adhesion and thus inflammatory activity in the vessel wall, maintains vascular permeability to nutrients, hormones, other macromolecules and leucocytes, inhibits platelet adhesion and aggregation by producing prostacyclin and NO. NO is a particularly important endothelium-derived mediator, because of its vasodilator, anti-platelet, anti-proliferative, permeability-decreasing and anti-inflammatory properties.[225] NO inhibits leukocyte adhesion and rolling as well as cytokine-induced expression of VCAM-1 (vascular cell adhesion molecule- 1) and MCP-1 (monocyte chemotactic protein-1),[226] effects that are at least in part attributable to inhibition of the transcription factor NF-κB (nuclear factor κB).[227] Injury to the endothelium can cause endothelial dysfunction. Dysfunction of the endothelium can be considered present when its properties, either in the basal state or after stimulation have changed in a way that is inappropriate with regard to the preservation of organ function. Endothelial dysfunction is thought to play an important role not only in the initiation of atherosclerosis, but also in its progression and clinical sequelae. Risk factors such as hypercholesterolaemia, dyslipidaemia, smoking and diabetes initiate atherosclerosis through endothelial activation and therefore endothelial dysfunction. Therefore endothelial activation can be conceptualized as a transducer of atherogenic risk factors.

A widely accepted theory states that insulin resistance is the primary abnormality that gives rise to Type II diabetes, hypertension and dyslipidaemia and that endothelial dysfunction merely represents the impact of hyperglycaemia and other features of the metabolic syndrome. An alternative concept is that endothelial dysfunction is at the heart of the metabolic syndrome. Endothelial dysfunction and obesity/insulin resistance frequently and closely coexist throughout the natural history of type 2 diabetes. Figure 2.12 illustrates this association as well as the various groups of individuals at risk for type 2 diabetes who have been found with these abnormalities Figure 2.13 integrates the potential mechanisms through which obesity, insulin resistance, and endothelial dysfunction are closely associated.

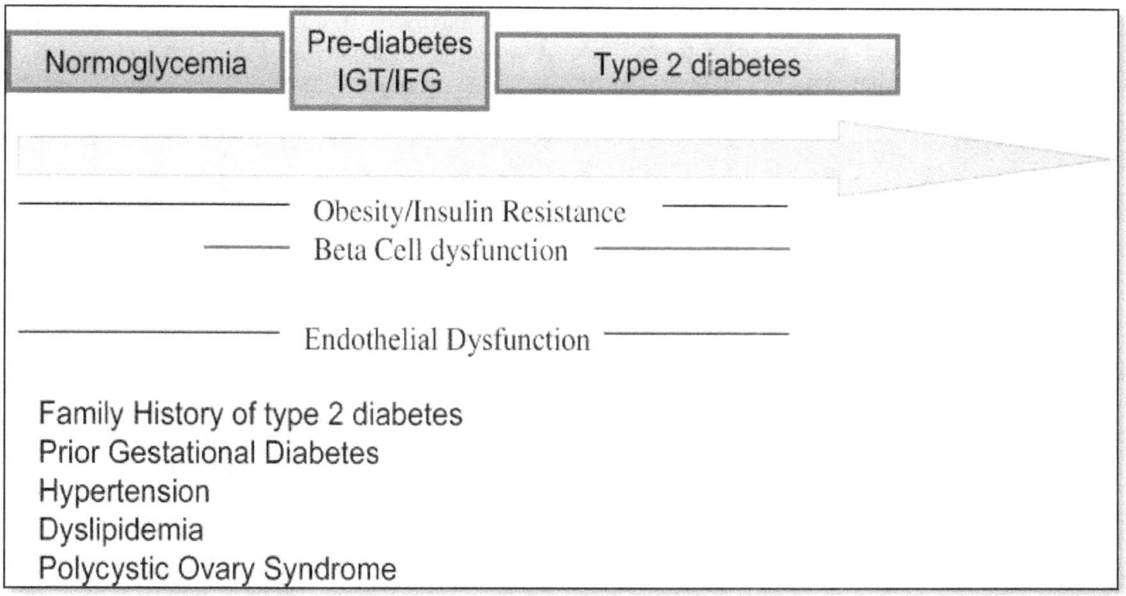

Figure 2.12 Obesity/insulin resistance and endothelial dysfunction closely coexist throughout the natural history of type 2 diabetes. They both are present in people with type 2 diabetes and in those in the prediabetic state, which includes people with IGT and/or IFG. Several groups at risk for type 2 diabetes, even at a stage of normal glucose homeostasis, have also been described with these abnormalities.[228]

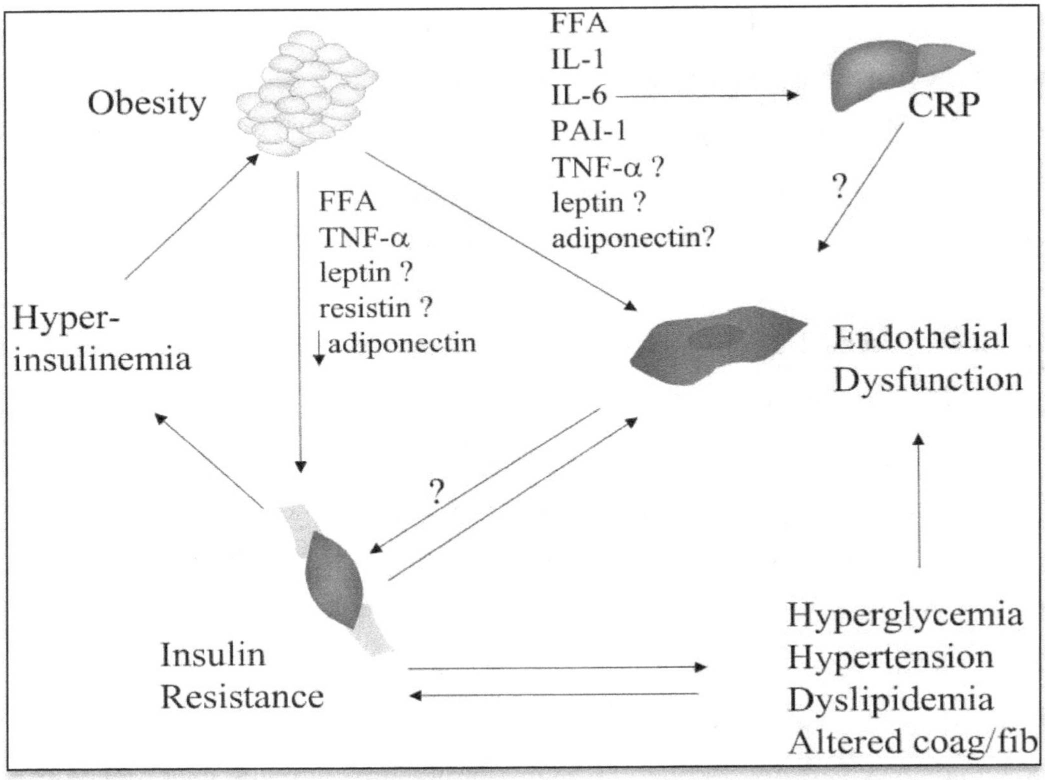

Figure 2.13 Mechanisms through which obesity, insulin resistance, and endothelial dysfunction are closely associated. Obesity leads to insulin resistance and endothelial dysfunction, mainly through fat-derived metabolic products, hormones, and cytokines (adipocytokines). Insulin resistance leads to endothelial dysfunction and may contribute to obesity. Insulin resistance is frequently associated with other abnormalities that can affect endothelial function, such as hyperglycemia, hypertension, dyslipidemia, and altered coagulation/fibrinolysis. Endothelial dysfunction may also favor insulin resistance. [228]

The mechanisms by which insulin resistance leads to endothelial dysfunction are certainly multiple and complex. All major abnormalities that are part of the insulin resistance syndrome, such as hyperglycemia, hypertension, dyslipidemia and altered coagulation/fibrinolysis, are directly and independently linked to endothelial dysfunction Insulin is known to have a direct vasodilatory effect mediated through stimulation of nitric oxide production in endothelial cells. In the insulin-resistant state, the ability of insulin to stimulate nitric oxide production in the endothelium is diminished. [228]

People with type 2 diabetes almost invariably have abnormal endothelial function as determined by the assessment of vascular reactivity and/or by the measurement of plasma markers of endothelial activation, coagulation/fibrinolysis, or inflammation. They have been consistently people with type 2 diabetes almost invariably have abnormal endothelial function as determined by the assessment of vascular reactivity and/or by the measurement of plasma markers of endothelial activation, coagulation/fibrinolysis, or inflammation. They have been consistently found to have abnormal small and large vessel reactivity for both endothelium-dependent and -independent vasodilatory pathways, demonstrating that there is not only a reduction in nitric oxide production in diabetes but also a decreased response to its effect in vascular smooth muscle cells.[229, 230] These abnormalities are often the result of hyperglycemia, hypertension, dyslipidemia, altered fibrinolysis, and obesity/insulin resistance.[231]

2.12.1 Nitric Oxide

Nitric oxide (NO) is a small, reactive molecule that is an extraordinarily important bioregulator. Although NO was first characterized more than 200 years ago by Joseph Priestly, its functions as a biomessenger were not recognized until the mid-1980s. NO is now recognized as a ubiquitous biomessenger, existing in a wide variety of organisms. In mammals, NO is used for a number of intercellular and intra-cellular signaling functions including blood-vessel dilation, neuronal signal transmission, cytotoxicity against pathogens and tumors, coordination of heart rhythm and the regulation of cellular respiration activity. Endogenous NO is produced by the heme-containing metalloenzyme nitric oxide synthase (NOS; EC 1.14.13.39).[232-234] There are several NOS isoforms and these are homodimers with each monomer containing binding sites for NADPH, flavin mononucleotide (FMN), flavin adenine dinucleotide (FAD), calmodulin (CaM), tetrahydrobiopterin (H4B) and a heme group. Catalytic activity assays reveal that H4B is an especially important NOS cofactor serving both allosteric and redox functions. The X-ray crystal structures of substrate bound NOS show that both the substrate and H4B are bound at the heme site with a substantial network of hydrogen bonds. A cysteinyl thiolate-ligated heme group reminiscent of cytochrome P450 is the redox active center that mediates the NOS reactions. NOS catalyzes the two step, five-electron oxidation of L-arginine (L-Arg) via *N*-hydroxy-L-arginine (NHA) to citrulline and NO. [235,236] The first step of the NOS reaction recruits two NADPH derived reducing equivalents to furnish the two-electron oxidation of L-Arg to NHA by $O_2^{\bullet}$.[237,238] The second step of the NOS reaction is unusual because it furnishes the three-electron aerobic oxidation of NHA to NO and citrulline by consuming only one NADPH-derived reducing equivalent as shown in figure 2.14.[239,240]

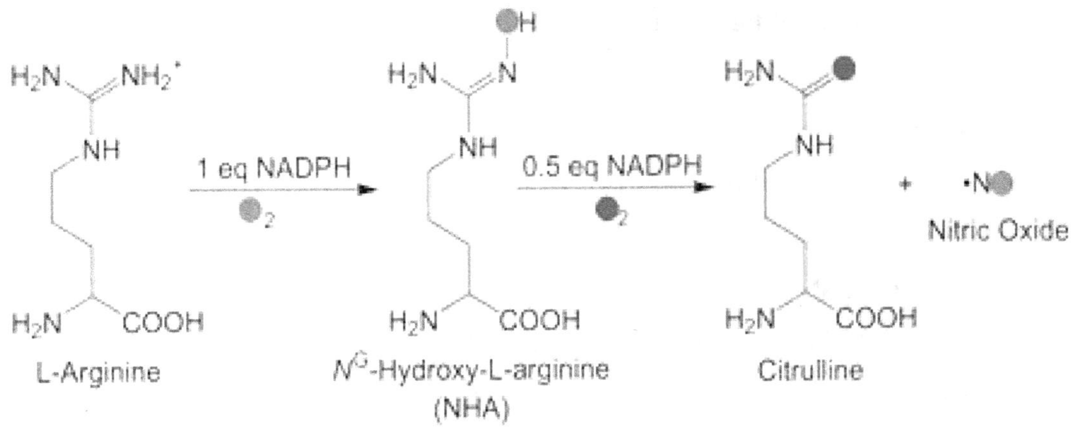

Figure 2.14 Nitric oxide synthase: models and mechanisms.[241]

2.12.2 Nitric oxide synthase enzyme

The catalytically active NOS is a homodimer of hemoprotein. The molecular weight of the monomer ranges from 110 kDa to 160 kDa depending on the isoform.[242,243] Each NOS monomer contains an oxidase domain at its amino-terminal end and a reductase domain at its carboxy terminal end. The NOS reductase domain contains binding sites for the redox cofactors NADPH, FMN, FAD and CaM. Two reducing equivalents are transferred from NADPH, via FMN and FAD, to the oxidase domain under the control of the Ca^{2+}/CaM complex. The oxidase domain tightly binds one equivalent each of H4B and a cysteinyl thiolate-ligated heme group, which is the reactive center of the oxidation reaction shown in figure 2.15.

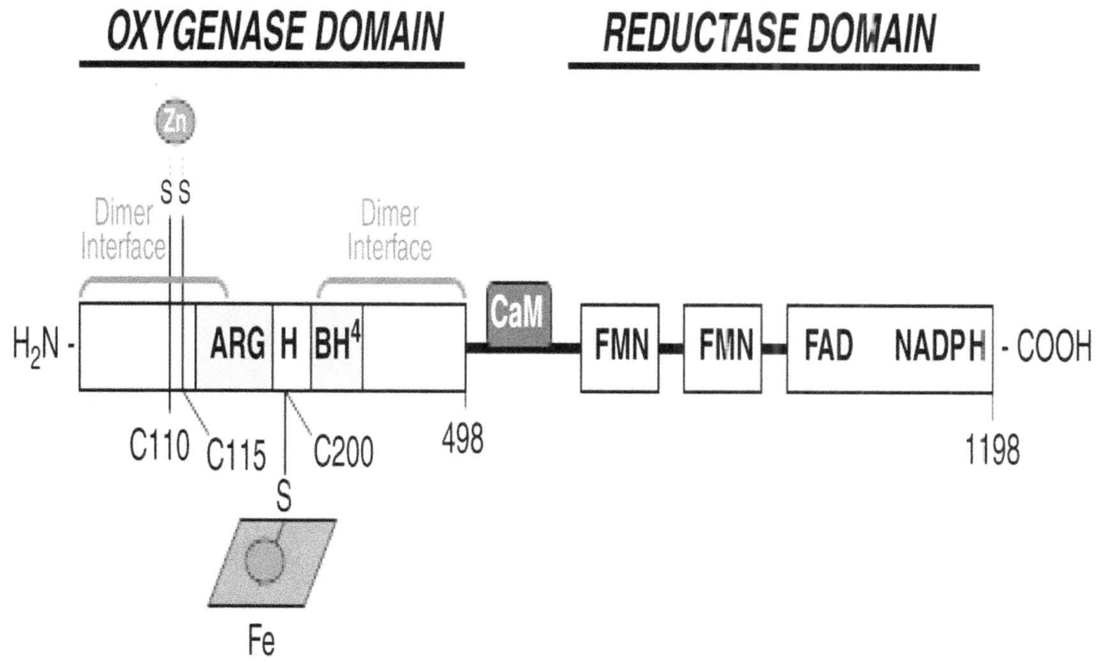

Figure 2.15 Domain structure of nitric oxide synthases (NOS). Binding sites for L-arginine (ARG), heme (H), tetrahydrobiopterin (BH4), and calmodulin (CaM) are indicated. The zinc tetrathiolate cluster is shown. Numbers indicate amino acid positions in murine inducible (i) NOS. FMN, flavin mononucleotide [244]

2.12.3 Isoenzymes of Nitric Oxide Synthase (NOS)[245]

NOS are a heme-containing enzyme with a sequence similarity to cytochrome P-450 reductase. Several isoforms of NOS are now known to exist, two of which are constitutive and one of which is inducible by immunological stimuli.[245]

i) The constitutive NOS (cNOS): It was first discovered in the vascular endothelium has been designated as eNOS.

ii) Neuronal NOS: The NOS present in the brain, spinal cord and peripheral nervous system is termed nNOS.

ii) The inducible NOS: The form of NOS induced by immunological or inflammatory stimuli is known as iNOS.

A comparison of the properties of these three major isoforms of the enzyme is shown in table 2.5.

Table 2.5 : Isoform of Nitric Oxide Synthase [245]

Isoforms of Nitric Oxide Synthase			
	Endothelial NOS (eNOS) [a] (Type III NOS, NOS-3)	Neuronal NOS (nNOS) [a] (Type I NOS, NOS-1, bNOS)	Inducible NOS (iNOS) [a] (Type II NOS, NOS-2, macNOS, hepNOS)
Primary regulation	Ca21/calmodulin	Ca21/calmodulin	Gene expression
Subcellular location	Membrane >> cytosol	Cytosol?	Cytosol >> membrane
NO output[b]	Low (pmolar)	Low (pmolar)	High (mmolar)
Function	Cell signaling	Cell signaling	Cytotoxic Cytostatic Cytoprotective

The complementary deoxyribonucleic acids for all three of these isoforms have been cloned from several species, including humans.[246-248] This has revealed further differences among them, which are shown in table 2.6. Knockout mice have been generated for each of the three NOS isoforms and have provided useful information concerning the role of each isoform and the effects of its deletion in whole animals.[249] The apparent association of the three NOS isoenzymes with the endothelium, neurons and inducibility (table 2.5) is an oversimplification. For example, eNOS is located not only in the vascular endothelial cells but also in platelets,[250] and in certain neuronal populations in the brain.[251] whereas nNOS has been found in the epithelium of the bronchi and trachea, as well as in skeletal muscle.[252] Furthermore, some differences have been identified between iNOSs obtained from different tissues within the same species. In addition, the constitutive eNOS can be induced in certain situations such as during chronic exercise,[253] or during pregnancy, when both eNOS and iNOS are induced,[254] whereas iNOS appears to be present constitutively in some tissues, including human bronchial epithelium,[252] rat kidney and some fetal tissues.[255]

2.12.4 Chromosomal location of eNOS gene:

eNOs gene is located on Chromosome 7, figure 2.17 shows the location of this eNOS gene,[256]

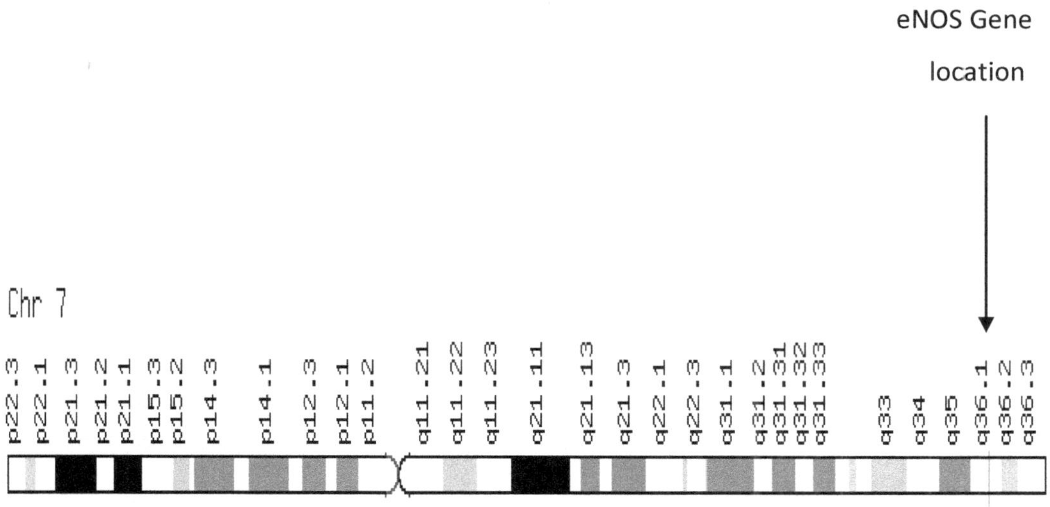

Entrez Gene cytogenetic band: 7q36 Ensembl cytogenetic and: 7q36.1 HGNC cytogenetic band: 7q36

Figure 2.16 Location of eNOS gene on chromosome 7

Table 2.6: Molecular biology of human isoenzymes of nitric oxide synthase

Molecular biology of human NOS isoforms[a]			
Gene	eNOS	nNOS	iNOS
Location	7q35–7q36	12q24.2	17cen–17q11.2
Structure	26 exons	29 exons	27 exons
Size	21 kb	160 kb	37 kb
Peptide	1203 aa	1434 aa	1153 aa
(deduced)	133 kDa	160 kDa	131 kDa

a Examples of accession numbers: eNOS - M95296; nNOS - U17327; iNOS - X73029.

2.12.5 Regulation of eNOS activity and gene expression

eNOS gene was considered constitutively expressed. The eNOS gene promoter was cloned and it was demonstrated that it has a complex regulatory mechanisms of gene expression. Like the promoter of constitutively expressed genes, the eNOS gene promoter does not contain the TATA box sequence. However, it has multiple cis-regulatory DNA sequences, including CCAT box, Sp1 sites, GATA motifs, CACCC box, AP-1 and AP-2 sites, p53 binding sites, and NF-1 elements, in addition to sequences responsive to sterol elements and shear stress.[257]

The complexity of the eNOS post-transcriptional regulatory mechanisms has been considered in the dimerization of the protein subunits and in the role of the protein caveolin in the formation of the caveolar structure.[258] Endothelial NOS is located inside the caveola and is kept in a less active state via its interaction with caveolin-126. The mechanisms migration from the cell membrane to the Golgi apparatus and Akt, PKA and AMPK-kinase-dependent phosphorylations are described as responsible for the activation of this NOS isoform. Thus, the interaction keeps eNOS inactive and calmodulin acts directly on the competition with caveolin in the promotion of the calcium-dependent enzyme activation.[259] Endothelial NOS activity is well known in the vascular environment and is regulated by six mechanisms after its translation: lipid inclusion; calcium/modulin-dependent mechanism; direct protein-protein interactions; several phosphorylations; glycosylation; and substrate and cofactor availability. Thus, eNOS can interact with several proteins in their "less active" or "more active" states.

The main eNOS activation mechanism occurs via phosphorylation of the amino acid serine at position 1.177 [256, 260] by the Akt-kinase enzyme (or kinase protein B), which increases eNOS sensitivity to baseline Ca^{++}/calmodulin concentrations.[261] Tonic or phasic eNOS activation in response to blood flow is independent of Ca^{++} concentration changes and constitutes the shear stress. Dimmeler et al [262] demonstrated that the exchange of the serine1177/1179 residue for the amino acid alanine makes eNOS incapable of responding to phosphorylation and activation by the Akt enzyme, a phosphatidylinositol-3-kinase (PI-3K)-dependent pathway. Although phosphorylation of the serine1177 residue plays a pivotal role in eNOS enzyme activation, its regulation is known to be also dependent on the phosphorylation pattern of other currently well-characterized sites.[263] Phosphorylation of the serine 633 residue, located in the binding domain of flavine mononucleotide (FMN) also increases eNOS activity and seems to be particularly important to keep NO synthesis after serine1177 residue phosphorylation and activation by Ca++/calmodulin. Phosphorylated by protein kinase C (PKC), the threonin 495 residue interferes with the calmodulin binding domain, thus downregulating NO synthesis. Recent studies have evaluated the capacity of certain drugs, such as antioxidants and renin-angiotensin system blockers, to reduce endothelial dysfunction by means mechanisms of eNOS activation via phosphorylation of specific sites such as the serine1177 residue. This phosphorylation may be affected by the subcellular enzyme location, such as the caveolae, intercellular junctions, Golgi apparatus and cytosolic compartments, as well as by protein kinases and phosphatases associated with these structures.

Several diseases have been associated with abnormalities in NO biosynthesis. Population genetics studies have demonstrated a significant association of eNOS gene G894T polymorphism with coronary artery disease (CAD).[264-266] Among the cardiovascular dysfunctions, the G894T variant was also demonstrated to be associated with hypertension, type 2 diabetes and complications related to it.[267,268] The correlation between genotype and clinical phenotype varies both quantitatively and qualitatively and the inconsistent association between eNOS polymorphism and several clinical phenotypes is a phenomenon commonly observed in other phenotype-associated genes. Studies have demonstrated the association between the G894T variant of the eNOS gene and cardiovascular phenotypes leading to reduction in NO bioavailability, thus suggesting that eNOS transcribed from the mutant allele has an abnormal enzymatic activity.

eNOS polymorphisms and functional studies on the G894T variant Genotyped and sequenced in 1993 by Marsden et al,[257] (GenBank D26607), eNOS is located in chromosome 7q35- 36 and variations in its sequence have been described in the promoter, exons and introns. The gene (21-22 kbp) comprises 26 exons and 25 introns with 133 kDa. The polypeptide sequence generated contains 1,203 amino acids.[257] The existence of three single nucleotide polymorphisms (SNP) in the promoter, in non-binding locations of transcription factors has already been described in the literature.[269] SNP were found in introns 2, 11, 12, 18, 22 and 23,[270] and repeat sequence polymorphisms in introns 2, 4, 8, and 1338.[271] Of the polymorphisms found in exons 6 and 7, replacement of the nitrogenous base guanine for thymine (G→T) at position 894 located in exon 7 results in a change of the amino acid glutamate (GAG) to aspartate (GAT) at position 298 of the polypeptide sequence.[272] It is suggested that polymorphisms located in the gene promoter have an influence on mRNA transcription, whereas polymorphisms located in coding regions may result in changes in enzymatic activity.[273]

2.12.6 Association between the G894T variant of the eNOS gene and cardiovascular phenotypes

It is observed that there are many diseases have been associated with abnormalities in NO biosynthesis and many of these conditions are related to autonomic dysfunction. Many studies have demonstrated a significant association of eNOS gene G894T polymorphism with coronary artery disease (CAD) [264-266] and also with acetylcholine (Ach)-induced coronary vasospasm.[274] Among the cardiovascular dysfunctions, the G894T variant was also demonstrated to be associated with hypertension although this association had not been verified in other populations.[275,276]

The correlation between genotype and clinical phenotype varies both quantitatively and qualitatively and the inconsistent association between eNOS polymorphism and several clinical phenotypes is a phenomenon commonly observed in other phenotype-associated genes.[277] This inconsistency has been attributed to environmental factors, independent alleles, interaction between genes and variability in clinical phenotypes. The importance of environmental factors in the genesis of diseases is reflected in morbidity and mortality differences between genetically homogeneous groups, however with different life styles.

A genetic variation may not be relevant to a determined population, thus reflecting differences in the frequency of allele distribution. One example is the frequency of the allele G894T, which is significantly higher in white populations in comparison to the Japanese population.[278] The inconsistent association between eNOS polymorphism and vascular changes is still being attributed to the variation in eNOS distribution in different organs. Arteries of specific organs are subject to different hemodynamic pressures, determining the vessel wall response and subsequent degree of endothelial dysfunction.[277]

Philip et al [272] studied patients undergoing surgical revascularization and observed that vascular reactivity to phenylephrine (PE) infusion is influenced by the G894T variant of the eNOS gene. The dose-dependent vasoconstrictor response to PE was significantly greater for the TT and GT alleles in comparison to the homozygote GG group, thus indicating that vascular reactivity to vasoconstrictor drugs may be influenced by eNOS polymorphism in humans. The greater dose-dependent response of patients with the G894T variant suggests reduced NO biosynthesis. Systemic administration of *N*G-nitro-L-arginine methyl esther (L-NAME) leads to hypertension in humans.[279,280] According to Frandsenn et al [281] the administration of 4mg/kg of L-NAME in humans reduces eNOS activity by 67 percent. The possibly reduced NO bioavailability subsequent to eNOS gene variations is an important candidate to the susceptibility to the development of endothelial dysfunction.[271,274] Studies demonstrating the association between the G894T variant of the eNOS gene and cardiovascular phenotypes point to a possible reduction in

NO bioavailability, thus suggesting that eNOS transcribed from the mutant allele has an abnormal enzymatic activity.

From the extensive review of literature it may be said that metabolic syndrome can be considered as a pre-diabetic state. It is viewed as the primary abnormality that gives rise to dyslipidemia, essential hypertension, impaired fasting glucose and impaired glucose tolerance. It imparts fivefold risk for diabetes. Type 2 diabetes which is characterized by hyperglycemia targets several organs, their physiological functions and leads to disturbed glucose homeostasis. The primary target of hyperglycemia appears to be endothelial cells, leading to its damage. Endothelial dysfunction is also an important component of the metabolic or insulin resistance syndrome and this is demonstrated by inadequate vasodilatation and/or paradoxical vasoconstriction in coronary and peripheral arteries in response to stimuli that release nitric oxide (NO). Nitric oxide is synthesized by enzyme nitric oxide synthase which is coded by eNOS gene. A functional polymorphism in this gene may affect the activity of the enzyme thus its functions. The review of different original research articles covering anthropometry parametes, glycemic parametes, lipid parameters, nitric oxide, eNOS gene polymorphism in metabolic syndrome, type 2 diabetes and its complications are mentioned below.

Haffner et al (1999)[282] verified the theory that Subjects converting to type 2 diabetes mellitus have increased cardiovascular risk factors relative to nonconverters and whether these atherogenic changes in the prediabetic state are predominantly due to insulin resistance, decreased insulin secretion, or both. They examined this issue in the 7-years follow-up of the San Antonio Heart Study, in which 195 of 1734 subjects converted to type 2 diabetes. At baseline, converters had significantly higher body mass index, waist circumference, triglyceride concentration and blood pressure and lower HDL cholesterol than nonconverters. Their data suggested that atherogenic changes in the prediabetic state are mainly seen in insulin-resistant subjects.

Katsuki et al (2001)[283] investigated usefulness of HOMA-IR and its relationship with insulin resistance assessed by the hyperinsulinemic-euglycemic clamp for evaluating the clinical course of patients with type 2 diabetes. Univariate regression analysis performed by him disclosed a significant correlation between log transformed HOMA-IR and log-transformed clamp IR before and after treatment. He concluded that HOMA-IR may constitute a useful method not only for diagnosing insulin resistance, but also for follow-up during the treatment of patients with type 2 diabetes.

Laaksonen et al (2002)[284] assessed the sensitivity and specificity of the definitions of the metabolic syndrome for prevalent and incident diabetes mellitus in a Finnish population-based cohort of middle-aged men (*n* = 1,005) followed for 4 years since the late 1980s. Four definitions based on the WHO and NCEP recommendations were compared. All definitions identified persons at high risk for developing diabetes during the 4-year follow-up (odds ratios = 5.0–8.8). The WHO definition including waist-hip ratio > 0.90 or body mass index ≥ 30 kg/m^2 was the most sensitive (0.83 and 0.67) but least specific (0.78 and 0.80) in detecting the 47 prevalent and 51 incident cases of diabetes. The NCEP definition in which adiposity was defined as waist girth > 102 cm detected only 61% of prevalent and 41% of incident diabetes, although it was the most specific (0.89 and 0.90). According to his study it is found that, WHO definition was found more valid as judged by its relatively high sensitivity and specificity in predicting diabetes. The NCEP definition including waist > 102 cm also identifies persons at high risk for diabetes, but it is relatively insensitive in predicting diabetes.

Yokoyama et al (2003)[285] investigated whether the quantitative insulin sensitivity check index (QUICKI) and the reciprocal index of homeostasis model assessment (1/HOMA-IR) derived from fasting plasma glucose and insulin level are excellent surrogate indices of insulin resistance in both normal range–weight and moderately obese type 2 diabetic and healthy subjects. They reported that QUICKI and 1/HOMA-IR were highly correlated with clamp-IR, with comparable coefficients in both normal range–weight and moderately obese type 2 diabetic patients and non diabetic subjects. They conluded 1/HOMA IR can probably be applied clinically as surrogate indices in view of its convenience.

A cross-sectional study done by **Palaniappan et al** (2003)[286] included 5659 women and men aged 20 to 80 years, to investigate association between microalbuminuria and metabolic syndrome. The result of their study revealed that microalbuminuria was present in 7.8% of women and 5.0% of men. Log linear analysis revealed a significant association between the metabolic syndrome and microalbuminuria in both genders. They concluded that microalbuminuria and metabolic syndrome are associated in a large, nationally representative cohort. He reported that it was possibly due to early renal effects of hypertension and it may be useful to consider microalbuminuria as a component of the metabolic syndrome.

Monti et al (2003)[287] in their association study of eNOS gene with type 2 diabetes and insulin resistance, demonstrated a significant association between eNOS gene polymorphisms and type 2 diabetes. They suggested that eNOS gene polymorphism is a new genetic susceptibility factor for hyperinsulinemia, insulin resistance and type 2 diabetes.

Hoffmann et al (2005)[288] investigated the prevalence of two eNOS polymorphisms, the Glu298Asp variant on exon 7 and the 4a/b variable number of tandem repeats (VNTR) on intron 4 and their association with blood pressure (BP), NO production, salt sensitivity and cardiovascular risk factors in healthy Venezuelans. They reported that prevalence of both polymorphisms in Venezuelans was comparable to that described for Caucasians, but significantly different from that known for African-Americans and Japanese. The 4a/b genotype was associated with reduced levels of NO metabolites (25% decrease), larger BP lowering in response to salt restriction (9.0 vs 4.8 mmHg, Po0.05), greater prevalence of salt sensitivity (39% in 4a/b and 27% in 4b/b; Po0.05) and with higher LDL-cholesterol levels. The Glu298T polymorphism did not affect NO production, nor it was associated with salt sensitivity. Glu298Asp polymorphism was positively associated with higher weight, triglycerides and LDL-cholesterol. Neither polymorphism was associated with changes in fasting or post load serum glucose, BP, obesity and albuminuria. They concluded that prevalence of eNOS polymorphisms is strongly determined by ethnic factors. The 4a/b gene polymorphism could be a genetic susceptibility factor for the BP response to salt intake and for the genetic control of NO production. The reduced NO production in subjects with the 4a/b genotype may be responsible for the increased sensitivity of their BP to salt.

Lu et al. (2006)[289] studied the clinical characteristics of hypertensive subjects with metabolic syndrome and its components in communities. They enrollled 5628 subjects aged over 20 years and measured height, weight, waist circumference (W) , hip circumference, systolic pressure (SP) , diastolic pressure (DP) , Lipid profile, fasting plasma glucose (FPG) , 2 h postprandial plasma glucose (2hPG) , fasting serum insulin (FIN) and 2 h postprandial serum insulin (2hIN). Body mass index (BMI), waist to hip ratio (WHR), and homeostatic model of insulin resistance index (HOMA-IR). They found that patients with isolated hypertension are rare in community-based population whereas about 85% of hypertensive patients had more than one metabolic disorders and more than half of were metabolic syndrome. The total body fat, levels of plasma glucose, serum insulin, total cholesterol and systolic blood pressure related with insulin resistance.

Nishevitha et al (2007)[290] determined the presence of genotype frequencies of Glu298Asp (G894T) single nucleotide polymorphism in the eNOS gene among south Indian Tamil speaking male population. Their result indicated that homozygous mutant genotype for eNOS polymorphism was absent/less common among south Indian male Tamil population of Asian race.

Ahluwalia et al (2008)[291] evaluated the association of the eNOS gene polymorphisms with type 2 diabetic nephropathy and observed significantly lower serum NO levels ($P < 0.05$) in diabetic patients carrying mutant allele of G894T. Comparatively higher levels of nitric oxide were found diabetic patients with and without nephropathy who were carrying the wild-type allele. Their results suggested that the eNOS gene locus is associated with diabetic nephropathy and the functional polymorphisms of eNOS gene(-894G > T) might lead to a decreased expression of eNOS gene.

Odeberg et al (2008)[292] carried out a population-based patient survey with 403 persons with T2DM and 799 healthy subjects from the population without diabetes or hypertension, They analysed the relation between T2DM, sex and the eNOS Asp298 allele versus the risk for AMI. Compared to wild-type non-diabetic subjects, all patients with T2DM had a significantly increased risk of AMI regardless of genotype. This risk was however markedly higher in patients with T2DM homozygous for the Asp298 allele (HR 7.20 [3.01–17.20], $p < 0.001$), independent of sex, BMI, systolic blood pressure, serum triglycerides, HDL -cholesterol, current smoking and leisure time physical activity. The pattern seemed stronger in diabetic women than in diabetic men. They suggested a synergistic effect of the eNOS Asp[298] allele and diabetes. They state that eNOS polymorphism as an important pathological bottleneck for cardiovascular disease in patients with T2DM. Also there is strong independent association between eNOS genotype and AMI.

Yang et al (2008)[293] investigated a possible association of polymorphism of the eNOS gene and essential hypertension in the Chinese Hui population; polymorphisms of rs2070744 (T>C), rs1799983 (G>T), rs1800780 (A>G), and rs3918181 (A>G) loci of the eNOS gene were examined. They found that polymorphisms of rs1799983 (G>T) and rs1800780 (A>G) in the eNOS gene produced potential genetic variants in essential hypertension in the Chinese Hui population, but not of rs2070744 (T>C) and rs3918181 (A>G). They reported that the rs1799983 (G>T) polymorphism is associated with essential hypertension susceptibility in this population. Individuals of this ethnicity are at risk for essential hypertension

Ianas et al. (2009)[294] studied G894T polymorphism of eNOS gene and the endocrine-metabolic changes in metabolic syndrome. They reported that the G894T polymorphism is associated with higher values of waist circumference, systolic and diastolic blood pressure, resistin, cortisol, 17-OHP, androstendione, PRL. The increased levels of cortisol, 17-OHP, androstendione, PRL in T allele carriers suggest a new genetic susceptibility component for hyperactivity of the hypothalamic-pituitary-adrenal axis leading to central adiposity and development of metabolic syndrome.

Chen et al.(2009)[295] compared body mass index, waist circumference, and waist-to-hip ratio as predictors of metabolic disorder in 1625 men and 1779 women, aged 35–64 as part of National Health Interview Survey. They reported that associations between anthropometric measures and the metabolic disorders varied with comorbidity, gender, age groups, and smoking status. They also mentioned that Waist-to-hip ratio was the best predictor for diabetes mellitus, especially for participants aged 45–64yrs.

Kaur et al (2010)[296] estimated the prevalence of the metabolic syndrome by a cross sectional survey and determined the association of risk factors with the metabolic syndrome in an urban industrial male population in Chennai, India. The total numbers of male subjects included in the study were 1077. The prevalence of the metabolic syndrome was 41.3% and 51.4% using IDF and AHA/NHLBI criteria respectively. Thus, the prevalence of the metabolic syndrome was very high in selected industrial population in south India.

Tangvarasittichai et al (2010)[297] analyzed the association of lipoprotein ratios with the homeostasis model assessment of insulin resistance (HOMA-IR) in type 2 diabetes mellitus (T2DM) patients (92) and 40 age matched healthy controls. All lipoprotein ratios, lipid profile, blood pressure and WC were significantly higher in type 2 diabetic patients as compared to healthy controls ($P<0.05$). TC/HDL-C ratio, TG/HDL-C ratio, non-HDL-C, WC, TG, and TC were significantly correlated with HOMA-IR ($P<0.05$) as obtained by Spearman correlation analysis.

Tessari et al (2010)[298] measured fractional (FSR) and absolute (ASR) synthesis rate of the nitric oxide rates in type 2 diabetic patients with diabetic nephropathy and in control subjects after infusion of L-[15N2-guanidino]-arginine. The study was conducted both before and after an euglycemic hyperinsulinemic (1,000 –1,200 pmol/l) clamp. They found that in type 2 diabetic patients and those with nephropathy, intravascular NOx synthesis from arginine is decreased under both basal and hyperinsulinemic states. This defect extends the concept of insulin resistance to altered NO metabolism.

Hsu et al (2011)[299]studied ssociation between insulin resistance and microalbuminuria in type 2 diabetes in cross-sectional study. They enrolled 738 normoalbuminuric type 2 diabetic subjects, aged 56.66±9.0 years and demonstrated increasing incidence rate ratios for those in the 2nd, 3rd and highest quartiles of HOMA IR, suggesting that insulin resistance could significantly predict development of microalbuminuria in type 2 diabetic patients.

Rani et al (2011)[300] conducted a population-based cross sectional study in cohort of 1414 subjects with type 2 diabetes from Chennai metropolis. All the subjects were subjected to eye examination and graded for diabetic retinopathy using Early Treatment Diabetic Retinopathy Study scales. The prevalence of microalbuminuria in the study subjects was 15.9% (226/1414) and that of macroalbuminuria, 2.7% (38/1414). Individuals with macroalbuminuria in comparison to micro- or normoalbuminuria showed a greater prevalence of DR (60.5% vs. 31.0% vs. 14.1%, $p < 0.001$), and also a greater severity of the disease (60.9% vs. 21.4 vs. 9.9, $p < 0.001$).

Shrama et al (2011) [301] carried out populations based cross sectional study including 150 diabetic patients. They investigated the association of dyslipidemia with diabetic retinopathic complications. It was demonstrated that prevalence of dyslipidemia was high in diabetic population; low levels of HDL along with fasting blood sugar level and duration of disease in type II diabetes are significantly associated with retinopathic complications.

Gosh et al (2011)[302] compared serum nitric oxide level among type 2 diabetic patients and healthy controls of Sikkim along with other biochemical parameters. They observed that serum NO was significantly low in diabetic participants as compared to control, along with difference in other biochemical parameters.

Mehta et al (2011)[303] performed cross sectional analyses of white subjects in two community based studies (N= 611 type 2 diabetic subjects, N= 803 non-diabetic subjects) using multivariate analysis of traditional risk factors and then adding MS, ApoB and homeostatic model assessment for insulin resistance (HOMA-IR). They stated that insulin resistance estimates add value to MS and Apo B in predicting coronary artery calcification scores in diabetes. They warranted that evaluation Apo B in diabetic patients will help in identifying of higher risk for atherosclerotic cardiovascular disease.

Angeline et al (2011)[5] studied association of endothelial nitric oxide synthase (eNOS) G894T single-nucleotide polymorphism with type 2 diabetes mellitus in South Indian (Tamil) population. They demonstrated that there is an association of endothelial nitric oxide synthase gene (G894T) polymorphism with diabetes mellitus among South Indians.

Shoukry et al (2012)[304] evaluated the association of three polymorphisms of the eNOS gene (894G>T, -786T>C, and 27-bp-VNTR) with the risk of diabetic nephropathy among type 2 diabetic patients. The C allele for -786T>C and the T allele for 894G>T were significantly more frequent in diabetics with nephropathy than in diabetics without nephropathy (p<0.001; odds ratio [OR] and 95 percent confidence interval [CI] for the C allele=1.64 [1.24-2.17] and p<0.001; OR and 95percent CI=1.7 [1.27-2.26] for the T allele. Their results suggested that the eNOS polymorphisms might represent genetic determinants for developing diabetic nephropathy in type 2 diabetic Egyptians.

Shah et al (2013)[305] examined the association of eNOS gene polymorphism in type 2 diabetic patients (T2DM) with and without diabetic nephropathy and observed that eNOS genotype carrying 'aa' genotype of 27VNTR (a/b) and eNOS haplotype C-a-G and C-a-T (allele of T-786C, 27VNTR a/b and G894T) carrying 'a' allele of 27VNTR (a/b) were associated with increased risk of diabetic nephropathy. They also observed lower serum NO levels in type 2 daibetic subjects (both cases and controls) carrying ba+aa genotypes. Their findings suggested that 27VNTR (a/b) eNOS polymorphism carrying 'a' allele alone and in association with T-786C and G894T eNOS polymorphism is associated with increased risk of DNP in Asian Indians T2DM patients.

Julinar et al (2013)[306] carried out a cross sectional study to determine correlation between NO and HOMA-IR in obese adolescent and found significant correlation between NO level and HOMA IR.

Salazar et al. (2013)[307] compared the ability of an elevated triglyceride/high-density lipoprotein cholesterol (TG/HDL-C) ratio, using sex-specific cut-points, to identify insulin-resistant individuals within a population without known cardiac disease or diabetes. Measurements were made of waist circumference, systolic and diastolic blood pressure, fasting plasma glucose, fasting plasma insulin, plasma TG and plasma HDL-C concentrations in 1102 women and 464 men. These data were used to classify subjects as being insulin resistant (FPI concentration in the upper quartile) and having the MS or an elevated TG/HDL-C ratio (>2.5 and >3.5 for women and men, respectively). The sensitivity and specificity with which the two indices identified insulin-resistant subjects were similar and hence they suggested that that TG/HDL-C concentration ratio is as adequate as MS diagnosis to identify insulin-resistant subjects.

El Mofty et al (2013)[308] detected the relation between apolipoprotein B (apo B) and diabetic retinopathy (DR) in a cross sectional study which included 168 patients with types II diabetes mellitus. There was a significant increase in Apo B in subjects with retinopathy as compared to those without retinopathy ($p<0.001$) and fair *c*orrelation of Apo B levels to the grade of retinopathy.

Kundu et al (2013)[309] determined microalbuminuria levels in type 2 diabetics and correlated changes in microalbuminuria levels with glycosylated hemoglobin level and duration of diabetes. They enrolled equal number of male and female subjects of type 2 diabetics between age groups 30-60 years, without any complications and 50 healthy (male and female) subjects in the study. Urinary microalbumin, HbA1c levels were significantly higher in diabetic cases. Microalbumin levels were linearly correlated to the duration of diabetes and HbA1c.

Okita et al (2013)[310] investigated the validity of HOMA-IR for evaluating insulin sensitivity in patients with type 2 diabetes on insulin therapy. They carried the study in two arms. In the first arm of the study, 19 poorly controlled diabetic subjects were treated with insulin and underwent euglycemic clamp study. Then the relationship between insulin resistance index assessed by the clamp test (clamp-IR) and HOMA-IR was investigated in these subjects. Log transformed HOMA-IR correlated with log transformed M/I values derived from the standard euglycemic clamp. In the second arm of the study, they investigated the relationship between HOMA-IR and various clinical parameters in 156 patients with poorly controlled diabetes after glycemic control. Log transformed HOMA-IR correlated negatively anthropometric and biochemical parameters like BMI, waist circumference, lipid profile urinary CPR and log transformed insulinogenic index. Their results suggest that HOMA-IR is a useful test for the evaluation of insulin sensitivity even in patients with type 2 diabetes treated with insulin.

Chapter 3
Hypothesis

Hypothesis

Health of endothelium is maintained by the local nitric oxide produced by nitric oxide synthase enzyme which is coded by eNOS gene located on chromosome7q35-36. A single nucleotide polymorphism at 894 results in G $\longrightarrow$ T transversion might be functionally related.

3.1 Null Hypothesis:

eNOS gene polymorphism is related to occurrence of Metabolic Syndrome, Type 2 Diabetes and Type 2 Diabetes with complications. This polymorphism is associated with changes in serum nitric oxide, urinary microalbumin and homeostasis model of assessment of insulin resistance (HOMA IR) which are the markers of endothelial dysfunction as well as insulin resistance.

3.2 Alternate hypothesis:

eNOS gene polymorphism is independent of occurrence of Metabolic Syndrome, Type 2 Diabetes and Type 2 Diabetes with complications. This polymorphism is not associated with changes in serum nitric oxide, urinary microalbumin and homeostasis model of assessment of insulin resistance.

Chapter 4
Aim and objectives

4.1 AIM

To assess the extent of endothelial dysfunction in patients of metabolic syndrome, type 2 diabetes with and without complications and correlate the same with endothelial nitric oxide synthase (eNOS) gene polymorphism.

4.2 OBJECTIVES

1. To study the **anthropometric parameters** in control and study groups
    - Body weight
    - Height
    - Body mass index (BMI)
    - Waist to hip (WHR) Ratio

2. To study the **glycemic parameters** in control and study groups
    - Fasting plasma sugar (FPS)
    - Glycosylated hemoglobin (HBA1c)
    - Fasting Insulin levels
    - Insulin Resistance by (Homeostasis Model of Assessment) HOMA

3. To study **lipid parameters** in control and study groups
    - Lipid Profile [Total Cholesterol (TC), Triglycerides(TG), High density lipoprotein(HDL), low density lipoprotein (LDL), Very low density lipoprotein (VLDL)]
    - Apolipoprotein B (Apo B)

4. **Endothelial Parameters** in control and study groups
    - Blood Pressure
    - Serum Nitric oxide
    - Urinary Microalbumin

5. **Genetic Analysis** in control and study groups
    - Genetic polymorphism of eNOS gene
    - Functional Correlation

Chapter 5

Material and Methods

Material & Methods

In order to study endothelial dysfunction in metabolic syndrome patients and diabetic patients with and without complications and to find association of eNOS gene polymorphism with phenotype parameters

5.1 Study design

Observational, Prospective

5.1.1 Ethical Clearance

Ethical Clearance was obtained on 24th April 2012 by Institutional Ethics Review committee.

5.1.2 Place of Study

The present study was carried out in the Department of Biochemistry and Department of Medicine, MGM Medical College, Kamothe, Navi-Mumbai.

Patients attending the medical & diabetic OPD were screened for the metabolic syndrome as per National Cholesterol Education Programme Adult treatment panel III (NCEP ATP III criteria)[9] and for type 2 diabetes mellitus as per Indian council of Medical Research (ICMR) / World Health Organization (WHO) guidelines.[26]

5.1.3 Study Population

The study population consisted of:

- Healthy controls
- Subjects of Metabolic syndrome
- Patients of Type 2 Diabetes Mellitus
- Patients of Type 2 Diabetes Mellitus with complications.

5.1.4 Enrollment of Patients

The patients visiting Diabetic OPD of MGM Hospital were enrolled after obtaining their informed written consent.

5.1.5 Sampling technique

Purposive Sampling

5.1.6 Study procedure

The patients visiting Diabetic OPD were first explained about the study and its purpose. After obtaining informed consent they were enrolled in the study. The various anthropometric parameters were measured and blood samples were collected as per the protocol for biochemical and genetic analysis.

5.1.7 Sample Size

Sample size was calculated from the following formula.[311]

Necessary Sample Size = $[\text{Z-score} \times \text{StdDev}]^2 / [\text{margin of error}]^2$

$N = [1.96 \times 6.95]^2/[1]^2 = [13.62]^2$

$N = 185.55 = 186$

5.2 Study groups

GROUP I (Healthy Controls n= 50)

Fifty age & sex matched healthy controls were enrolled for the study.

Inclusion Criteria:

The enrolment in this group was age & sex matched, healthy individuals without any clinical history of diabetes.

Exclusion Criteria:

On enrollment and before the study of various parameters if the inclusion criteria changed then the individual was excluded from the study. Known cases of diabetes were excluded from the study.

GROUP II (Metabolic Syndrome n= 60)

Sixty individuals suffering from metabolic syndrome were included in this study.

Inclusion Criteria:

Individuals fulfilling the NCEP ATP III criteria for metabolic syndrome were included in this group. NCEP ATP III criteria are as follows. (The three risk factors considered for identification of metabolic syndrome were BP, FPG and TG).

NCEP ATP III criteria for diagnosis of metabolic syndrome[9]

Risk Factor	Defining Level
Abdominal obesity Men Women	**Waist circumference** >102 cm (>40 in) >88 cm (>35 in)
Triglycerides	≥150 mg/dl
HDL Cholesterol Men Women	 <40 mg/dl <50 mg/dl
Blood pressure	≥130/≥85 mmHg
Fasting glucose	≥110 mg/dl

Exclusion Criteria:

Individuals showing alteration in the NCEP ATP III criteria for metabolic syndrome after enrollment were excluded from this group. Patients not satisfying diagnostic criteria for metabolic syndrome were excluded.

GROUP III (Diabetic patients n =60)

Sixty known type 2 diabetic patients diagnosed since one year duration were included in this group.

Inclusion Criteria:

Patients fulfilling WHO/ICMR criteria for type 2 diabetes mellitus were included in the study.

Exclusion Criteria:

Patient not fulfilling or showing variation in WHO/ICMR criteria for type 2 diabetes mellitus after enrollment were excluded from the study. WHO Criteria for diagnosis of type 2 diabetes mellitus is as follows.

ICMR /WHO Criteria for diagnosis of type 2 diabetes mellitus[312, 313]

Diabetes	
Fasting plasma glucose	≥7.0mmol/l (126mg/dl)
2–h plasma glucose	≥11.1mmol/l (200mg/dl)
Impaired Glucose tolerance (IGT)	
Fasting plasma glucose	<7.0mmol/l (126mg/dl) &
2–h plasma glucose	≥7.8 and <11.1mmol/l (140mg/dl and 200mg/dl)
Impaired fasting Glucose (IFG)	
Fasting plasma glucose	6.1 to 6.9mmol/l 110mg/dl to 125mg/dl and 100 mg/dl to 126 mg/dl as per WHO
2–h plasma glucose*	≥7.8mmol/l (140mg/dl) and <11.1 mmols/l (200 mg/dl)

GROUP IV (Diabetic patients with secondary complication n =60)

Sixty diabetic patients with secondary complicationss were included in the study.

Inclusion Criteria

Type 2 diabetic patients with secondary complications like nephropathy, neuropathy, retinopathy & cardiovascular diseases were included in the study.

Exclusion Criteria:

Endocrinological disorders other than type 2 diabetes, patients having illness of major organs and or cancer were excluded. Patients suffering satisfying diagnostic criteria for metabolic syndrome and diabetic patients without complications were excluded.

5.2.1 Data collection

Demographic data such as name, sex, age as well as patient's clinical history such as duration of diabetes, family history were collected.

5.3 Laboratory analysis

5.3.1. Anthropometric parameters

A. Weight and height measurements

Weight of the participants was measured using the weighing scale. Weight was taken without shoes and in light clothing. Height was measured in centimeters with subjects standing straight. Body mass index (BMI) for each individual was calculated by dividing weight (kilograms) with height squared (meter). BMI of 23kg/m^2 and over was taken as obesity.

B. Waist to Hip Ratio (WHR)

Waist to Hip ratio was calculated on dividing waist circumference by hip circumference in centimeters.

C. Blood pressure measurements

Blood pressure was taken from the arm (brachial artery) for all participants by using digital sphygmomanometer. Blood pressure measurement was done in a sitting position with the arm supported

5.3.2 Glycemic Parameters

A. Estimation of plasma glucose by Glucose oxidase and peroxidase method (Kit by ERBA, Transasia) [314,315]

Principle

α-D-glucose in the sample is rapidly converted to the β-isomer by the action of mutarotase which then is oxidized to yield gluconic acid and hydrogen peroxide in the presence of Glucose oxidase. The enzyme peroxidase catalyses the oxidative coupling of 4-aminoantipyrine with phenol to yield a coloured quinonemine complex, with absorbance at 505 nm proportional to the concentration of glucose in sample.

Reaction

$$\alpha\text{-D-glucose} \xrightarrow{\text{Mutarotase}} \beta\text{-D-Glucose}$$

$$\beta\text{-D-Glucose} + H_2O + O_2 \xrightarrow{\text{Glucose oxidase}} \text{Gluconic acid} + H_2O_2$$

$$H_2O_2 + \text{phenol} + 4AAP \xrightarrow{\text{Peroxidase}} \text{Red dye} + 2\,H_2O$$

Reagents and Test Composition

S.No	Active ingredient	Concentration
1.	Tris-phosphate buffer	50 mmol/l
2.	Glucose oxidase	≥ 20000 U/l
3.	Peroxidase	≥ 800 U/l
4.	Mutarotase	≥ 1000 U/l
5.	Phenol	10 mmols/l

Calculation
The Glucose concentration was computed by auot analyzer.

Reference Interval:

Fasting Glucose: 70-110 mg/dl
Post prandial glucose: 90-130 mg/dl

B. Estimation of fasting plasma insulin by ELISA method[316,317] (Kit by Immunoshop, Catlog no. KAP 1251)

Principle:

ELISA is a solid phase enzyme amplified sensitivity immunoassay performed on breakable microtirerplates. The assay uses monoclonal antibodies directed against distinct epitomes of insulin. Sample reacts with the monoclonal antibody (MAb1) coated on microtiter well and with a monoclonal antibody (MAb2) labeled with horseradish peroxidase (HRP). After an incubation period allowing sandwich formation (MAb1-human insulin-MAb2-HRP) the microtitre is washed to remove unbound enzyme labeled antibody. Bound enzyme labeled antibody is measured through a chromogenic reaction. Chromogenic solution is added and incubated. The reaction is stopped by addition of stop solution. The amount of substrate turnover is determined colourimetrically by measuring the absorbance at 450 nm.

Reagents provided in the kit

S. No.	Reagent	96 test kit	Reconstitution
1.	Microtiter plate with 96 anti INS coated breakable wells	96 wells	Ready for use
2.	Conjugate: HRP labeled anti-INS in Tris HCL buffer with bovine serum albumin and thymol	1 vial of 6 ml	Ready for use
3.	Zero calibrator in human serum and thymol	1 vial lyophilized	Add 2 ml distilled water
4.	Calibrator N=1 to 5	5 vials lyophilized	Add 1 ml distilled water
5.	Wash Solution	1 vial of 10 ml	Dilute 200x with distilled water
6.	Controls-N=1or 2	2 vials lyophilized	Add 1 ml distilled water
7.	Chromogenic TMB Solution	1 vial of 12 ml	Ready for use
8.	Stop Solution HCL 1.0 N	1 vial of 12 ml	Ready for use

Reagent preparation

A. Calibrators:

The zero calibrator is reconstituted with 2.0 ml distilled water and other calibrators with 1.0 ml distilled water.

B. Controls: The controls are reconstituted with 1.0 ml distilled water.

C. Working wash solution: Adequate volume of working wash solution is prepared by adding 199 volumes of distilled water to 1 volume of wash solution (200x), working solution is homogenized by magnetic stirrer.

Procedure

1. The required numbers of wells per run were selected. The unused wells were resealed in the bag with a desicant and stored at 2-8^{0}C.
2. 50 ul of each calibrator, control and sample was pipetted into appropriate wells.
3. 50 ul of anti-INS HRP conjugate was added into all the wells.
4. All the wells were incubated for 30 minutes at room temperature.
5. The liquid was aspirated from all the wells.
6. The plate was washed 3 times by dispensing 0.4 ml of wash solution in to each well and aspirating the content of each well.
7. 100 ul of chromogenic solution was added into each well within 15 minutes after the washing step.
8. The microtiter plate was incubated for 15 minute at room temperature.
9. 100 ul of stop solution was added into each well.
10. The absorbance was read at 450 nm on ELISA analyzer and reader by tranasaia.

Calculation of result

1. The plate was read at 450 nm against a reference filter set at 650 nm

2. The OD of calibrators was plotted as a liner graph against corresponding concentrations of insulin.

3. The concentration of each control and sample was read by interpolation on the calibration curve.

Figure 5.1 Plasma Insulin kit by Immunoshop

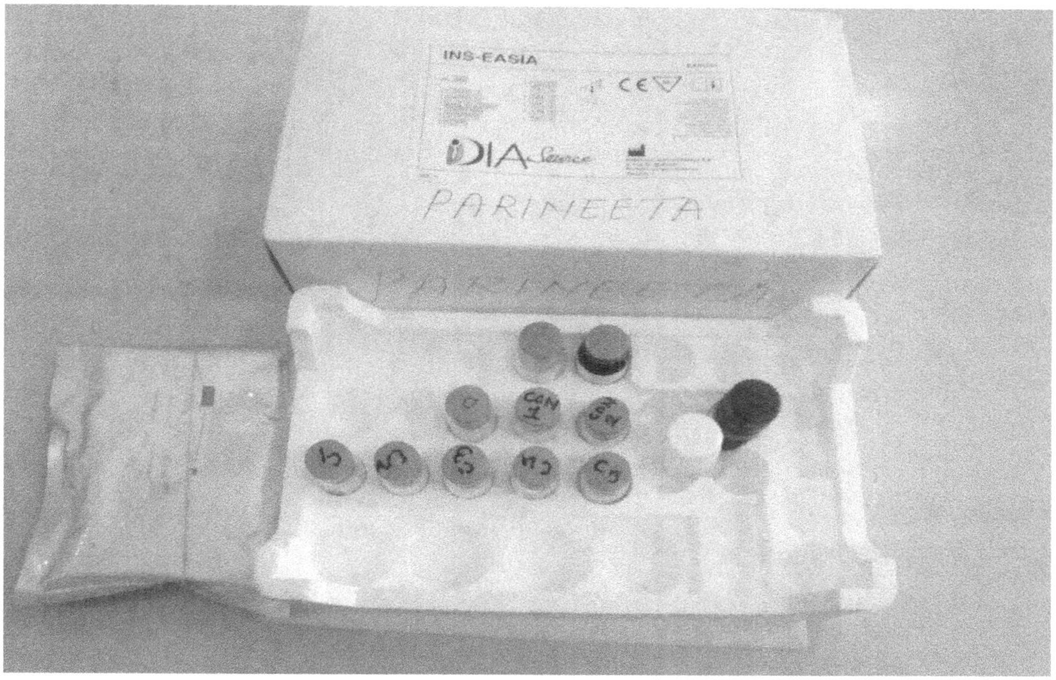

Figure 5.2a. ELISA assay for Insulin after addition of chromogenic solution.

Figure 5.2b. ELISA assay for Insulin after addition of stop solution.

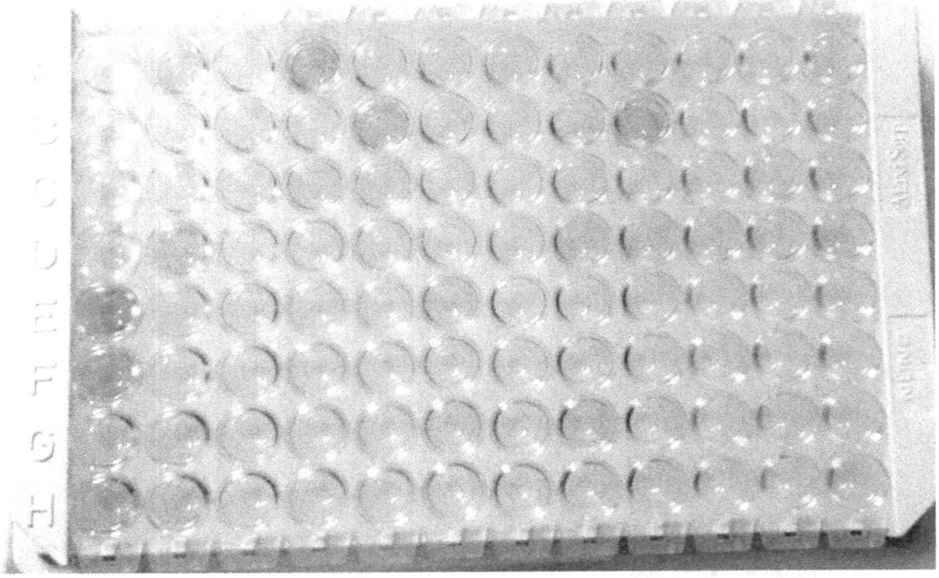

Calculation of HOMA IR:

The values of HOMA IR were calculated using following formula: [199]

Fasting plasma glucose in mg/dl X fasting plasma insulin in uIU/ml / 405

C. Estimation of HbA1c by antigen-antibody reaction[318, 319]

Total Hb and HbA1c in hemolysed blood bind with same affinity to particles in R1. The amount of binding is proportional to the relative concentration of both substances in the blood. Mouse anti human HbA1c monoclonal antibody (R2) binds to particle bound HbA1c. Goat anti mouse IgG polyclonal antibody (R3) interacts with monoclonal mouse anti-human HbA1c antibody and agglutination takes place. The measured absorbance is proportional to the HbA1c bound particles, which in turn is proportional to percentage of HbA1c in the sample.

Reagents

R1	Buffer	20 mmols/l
	Latex	0.14%
R2	Buffer	10 mmol/l
	Mouse anti human HbA1c monoclonal antibody	5.5 mg/dl
R3	Buffer	10 mmol/l
	Goat anti-mouse IgG polyclonal antibody	67 mg/dl
	stabilizer	

Reagent Preparation:

Reagents are ready to use.

Specimen: Whole blood collected with EDTA

Sample Preparation:

Hemolysing solution: 1000 ul
Sample/ Calibrator/Control: 20 ul
Mix and allow to stand for 5 min or until complete lysis is apparent.

Assay procedure

Sample or Calibrator	20 ul
Reagent 1	750 ul
Mix and incubate for 2 min., then add:	
Reagent 2	250 ul
Mix and incubate for 3 min., then add:	
Reagent 3	125 ul
Mix, read absorbance after exactly 2 min	

Calculation:

The concentration of HbA1c in unknown sample is derived from a calibration curve using appropriate mathematical model such as e.g. spline by autoanalyzer.

5.3.3 Lipid parameters

A. Estimation of total cholesterol by Cholesterol Esterase and Cholesterol Oxidase (CHO & POD) enzymatic method [320]

Principle

The Cholesterol reagent utilizes an enzymatic method to measure cholesterol in human serum and plasma. In this procedure cholesterol esters in sample are hydrolysed by cholesterol esterase (CHE). The free cholesterol produced is oxidised by cholesterol oxidase (CHO) to cholestene-3-one with the simultaneous production of hydrogen peroxide (H_2O_2), which oxidatively couples with 4-aminoantipyrine and phenol in the presence of peroxidase (POD) to yield a chromophore. The red quinoneimine dye formed can be measured spectrophotometrically at 540/600 nm as an increase in absorbance.

Reaction

$$\text{2 Cholesterol esters} + 2\,H_2O \xrightarrow{\text{CHE}} \text{2 Cholesterol} + \text{2 Fatty acids}$$

$$\text{2 Cholesterol} + 2\,O_2 \xrightarrow{\text{CHO}} \text{Cholestene-3-one} + 2\,H_2O_2$$

$$2\,H_2O_2 + \text{4-Aminoantipyrine} + \text{Phenol} \xrightarrow{\text{POD}} \text{Quinoneimine} + 4\,H_2O$$

Reagents and Test Composition

Final concentration of reactive ingredients:

1.	Phosphate buffer (pH 6.5)	103 mmol/L
2.	4-Aminoantipyrine	0.31 mmol/L
3.	Phenol	5.2 mmol/L
4.	Cholesterol esterase	0.2 kU/L (3.3 µkat/L)
5.	Cholesterol oxidase	≥ 0.2 kU/L (3.3 µkat/L)
6.	Peroxidase	≥ 10.0 kU/L (166.7 µkat/L)
7.	Preservative	

Calculation:
The total cholesterol concentration of each sample was computed by autonalyzer.

Reference Interval:

< 5.2 mmol/L (200 mg/dL)	Desirable
5.2 – 6.2 mmol/L (200 – 239 mg/dL)	Borderline High
≥ 6.2 mmol/L (240 mg/dL)	High

B. Estimation triglyceride by glycerol phosphate oxidase & peroxidase (GPO-POD) in serum by enzymatic method [321-323]

Principle

The triglyceride procedure is based on a series of coupled enzymatic reactions. The triglycerides in the sample are hydrolysed by a combination of microbial lipases to give glycerol and fatty acids. The glycerol is phosphorylated by adenosine triphosphate (ATP) in the presence of glycerol kinase (GK) to produce glycerol-3-phosphate. The glycerol-3-phosphate is oxidised by molecular oxygen in the presence of GPO (glycerol phosphate oxidase) to produce hydrogen peroxide (H_2O_2) and dihydroxyacetone phosphate. The formed H_2O_2 reacts with 4-aminophenazone and N,N-bis (4-sulfobutyl)-3,5-dimethylaniline, disodium salt (MADB) in the presence of peroxidase (POD) to produce a chromophore, which is read at 660/800nm. The increase in absorbance at 660/800nm is proportional to the triglyceride content of the sample.

Reaction

$$\text{Triglycerides} + 3\,H_2O \xrightarrow{\text{Lipase}} \text{Glycerol} + 3\ \text{Fatty acids}$$

$$\text{Glycerol} + \text{ATP} \xrightarrow{\text{GK},\ Mg^{2+}} \text{Glycerol-3-phosphate} + \text{ADP}$$

$$\text{Glycerol-3-phosphate} + O_2 \xrightarrow{\text{GPO}} \text{Dihydroxyacetone phosphate} + H_2O_2$$

$$H_2O_2 + 4\text{-AAP} + MADB \xrightarrow{POD} \text{Blue Dye} + OH^- + 3\,H_2O$$

Reagents and Test Composition

Final concentration of reactive ingredients:			
PIPES buffer (pH 7.5)	-50 mmol/L	Lipases	1.5 kU/L
Mg^{2+}	- 4.6 mmol/L	Glycerol kinase	5 kU/L
MADB	- 0.25 mmol/L	Peroxidase	0.98 kU/L
4-Aminoantipyrine	- 0.5 mmol/L	Ascorbate oxidase	1.48 kU/L
ATP	-1.4 mmol/L	Glycerol-3-phosoxidase	1.48 kU/L

Calculation:

The triglyceride concentration of each sample was computed by autoanalyzer.

Reference Intervals:

Normal	150 mg/dl (< 1.70 mmol/L)
Borderline high	150-199mg/dl (1.70 – 2.25 mmol/L)
High	200-499mg/dl (2.26 – 5.64 mmol/L)
Very high	≥ 500 mg/dl (5.65 mmol/L)

C. Estimation of serum HDL by Cholesterol Esterase and Choesterol Oxidase Immunoinhibition method (CHE and CHO immunoinhibition method) [324]

Principle

Anti human-β-lipoprotein antibody in R1 binds to lipoproteins other than HDL (LDL, VLDL and chylomicrons). The antigen-antibody complexes formed block enzyme reactions when R2 is added. HDL-cholesterol is quantified by the presence of an enzyme chromogen system.

Reaction

$$\text{LDL, VLDL and chylomicrons} \xrightarrow{\text{Anti human - } \beta \text{ - lipoprotein antibody}} \text{Antigen–Antibody complexes}$$

$$\text{HDL-cholesterol} + H_2O + O_2 \xrightarrow{\text{CHE and CHO}} \text{Cholest-4-en-3-one} + \text{Fatty acids} + H_2O_2$$

$$H_2O_2 + 4 - AA + F\text{–DAOS} \xrightarrow{\text{POD}} \text{Blue dye+} + F^- + 2H_2O$$

Reagents and Test Composition

S.No	Reagent	Volume
1.	Anti human-β-lipoprotein antibody	**variable**
2.	Cholesterol esterase (CHE)	0.8 IU/mL
3.	Cholesterol oxidase (CHO)	4.4 IU/mL
4.	Peroxidase (POD)	1.7 IU/mL
5.	Ascorbate oxidase	2.0 IU/mL
6.	Good's buffer (pH 7.0)	30 mmol/L
7.	N-Ethyl – N - (2-hydroxy-3-sulfopropyl) - 3.5– dimethoxy – 4 fluoroaniline (F–DAOS) 4-Aminoantipyrine	0.20 mmol/L
8.	Preservative	

Calculation:

Results were automatically computed by auto analyzers.

Reference Interval:

<1.03 mmol/L (<40 mg/dL)	Low HDL-cholesterol (major risk factor for coronary heart disease)
≥ 1.55 mmol/L (≥ 60 mg/dL)	High HDL-cholesterol ("negative" risk factor for coronary heart disease)

D.Calculation of LDL cholesterol and VLDL cholesterol

The LDL cholesterol and VLDL cholesterol were calculated by using the Friedwald's formula.[325]

Calculation:-

LDL-Cholesterol (mg/dl) = TC - HDL - TG/5.0

VLDL-Cholesterol (mg/dl) =Triglycerides/5

E. Estimation of Apolipoprotein B in serum by Immunoturbidometry method (ERBA) [326,327]

Apolipoprotein B is the main protein component of LDL (Low density lipoprotein). Apolipoprotein B is necessary for the reaction with LDL receptors in the liver and on cell walls and is thus involved in transporting cholesterol from the liver to the blood vessel cells. Elevated levels of Apolipoprotein B frequently found in atherosclerosis.

Principle

Measurement of antigen-antibody reaction by end point method.

Reagents and test composition

A. Antiserum:
- Phosphate Buffer Saline
- Polyclonal goat antihuman Apolipoprotein B
- Sodium azide
B. Buffer:

- Phosphate Buffer Saline
- Polyehylene glycol
- Detergent

- Sodium azide
C. Calibrator
- Buffered human plasma, lyophilized and stabilized for 1 ml
- 0.09% Sodium azide as preservative

Test protocol

Content	Volume
Serum sample	40 μl
Saline 9 g/l	900 μl
Antiserum	80 μl

It was incubated for 5 minutes and optical density was read at 340 nm.

Calculation:
The Apo B concentration of each sample was computed by autoanalyzer.

5.3.4 Endothelial parameter

A.Estimation of Microalbumin by Immuno turbidimetric method [328-330]

Principles

When a sample is mixed with R1 buffer and R2 antiserum solution human albumin reacts specifically with anti human albumin antibodies, to yield insoluble aggregates. The absorbance of these aggregates is proportional to the albumin concentration in the sample.

Reagents:

Tris buffer (pH 7.6)
Goat antihuman albumin antibodies
Preservative

Reagents preparation: All reagents were ready to use.

Test protocol

Content	Volume
Urine sample	1.6 µl
R1	15 µl
R2	10 µl

The test was read at 340 nm.

Calculation:
The urinary microalbumin concentration of each sample was computed by autoanalyzer.

B.Estimation of serum nitric oxide by modified Griess method [331,332]

Principle:-

Nitrate the stable product of nitric oxide is reduced to nitrite by cadmium reduction method after deproteinisation of sample. The nitrite produced is determined by diazotization of sulphanilamide and coupling to naphthylene ethylene diamine. The color complex produced is measured at 540 nm.

Reagents:-

1. Cadmium granules
2. Glycine, NaOH buffer (pH 9.7):- 7.5 gm of glycine was dissolved in 200ml deionised water then the pH was adjusted to 9.7 by 2 M NaOH and diluted to 500 ml with deionised water.
3. Sulphanilamide: - 2.5 gm of sulphanilamide was dissolved in 250 ml of warm 3 mol / l HCL and allowed to cool. This reagent was stable for one year at RT.
4. N- naphthylene ethylene diamine**:** - 50 mg of N - naphthylene ethylene diamine was dissolved in deionised water and the volume was adjusted to 250ml.
5. Preparation of the Griess Reagent

   Equal volumes of *N*-(1-naphthyl) ethylenediamine and sulfanilic acid were mixed to form the Griess Reagent.

6. Standard sodium nitrite solution: - Stock Std (0.1 mol/ l) 690 mg of sodium nitrite was dissolved in 100ml of 10 mmol / L sodium borate solution
7. Working standard solution (10 μ mol / L):- 10 μl of stock $NaNO_2$ was diluted to 100 ml with 10 m mol/ l sodium borate solution.
8. Zinc sulphate solution (75 m mol/L):- 2.15 gms of $ZnSO_4$ was dissolved in 100 ml D/W.
9. Sodium hydroxide (55 m mol/ L):- 220 mg of NaOH was dissolved in 100 ml D/W.
10. Sulphuric acid (0.1mol/L):- For storage of cadmium granules.
11. Copper sulphate (CuSO4):- 125 mg of $CuSO_4$ was dissolved in 100 ml of glycine sodium hydroxide buffer.
12. 2M Sodium hydroxide: - 8 gm of NaOH was dissolved in 100 ml D/W.
13. 3 mmol HCl: - 27.35 ml of NaOH was diluted to 250 ml D/W.
14. 10 mmol sodium borate:-1.90 gm od sodium Borate was dissolved in 500 ml D/W.

Procedure:-

Deproteinization: - Serum sample: - 0.5 ml

75 mmol/L ZnSO4:- 2.0 ml

55 mmol/ L NaOH:- 2.5ml

The tube was mixed well and centrifuged for 10 minutes. The clear supernatant was treated as protein free filtrate.

Activation of cadmium granules

Stored cadmium granules in 0.1 mol/ L sulphuric acid was rinsed thrice with deionised water at the time of assay. The cadmium granules were swirled in 5 m mol/ L $CuSO_4$ for 1-2 minutes (4-5ml).Then $CuSO_4$ was drained (4-5ml) and the Cu- coated granules were washed by glycine NaOH buffer. These activated granules were used within 10 minutes after activation.After utilisation, cadmium granules were drained and washed with deionised water and stored in 0.1 mol/L H_2SO_4. The same procedure of activation was followed for next time.

Assay procedure

1- 2 grams of activated granules were added to each flask containing 500 microlitre of protein free sample or standard (water was added and volume was adjusted to 2.0 ml). The contents of flasks were mixed to swirl the granules. After 20.0 minutes, 1.0 ml of above diluted solution from respective flasks was pipetted out in clean dry test tubes, and 1 ml of griess reagent was added. Tubes were mixed well and reading was taken at 540 nm after 20 min.

Calculation

$$\text{Serum Nitric Oxide } (\mu mol/L) = \frac{ODT - ODB}{ODS - ODB} \times 100$$

5.3.5 Genetic Analysis

A. Extraction of human genomic DNA from blood containing EDTA.[333,334]

SIGMA-ALDRICH GenElute Blood Genomic DNA kit was used for DNA extraction from the whole blood (Cat no NA2010).

Sigma GenElute Blood Genomic DNA Kit blood mini kits are designed for rapid purification of an average of 4 - 10 µg of total DNA from 200 µl of whole human blood.

Procedure:-

1. Collection of Blood

Whole blood was collected in EDTA tube which was equilibrated to room temperature before beginning the preparation.

2. Preparation of Blood

20 µl of Proteinase K solution was added into 1.5 ml microcentrifuge tube. Then 200 µl of whole blood sample was added to the tube. It was vortexed to ensure through mixing of the enzyme.

3. **Lysing of Cells**

200 µl of Lysis Solution C was added to the sample and vortexed thoroughly for 15 second. The homogenous mixture was incubated at 55°C for 10 minutes for efficient lysis.

4. **Preparation of Column**

500 µl of the column preparation solution was added to pre-assembled Genelute miniprep binding column and centrifuged at 13,500 rpm for 1 minute.

5. **Preparation for Binding**

200 µl of ethanol (95-100%) was added to the lysate from step 3 and mixed thoroughly by vortexing for 5-10 seconds.

6. **Loading of Lysate**

The entire lysate was transferred to the treated column from step 4 and centrifuged at 10,100 rpm for 1 minute. Collection tube was discarded and column was placed in another collection tube.

7. **First Wash**

500 µl of Pre wash Solution was added to the column and centrifuged at 10,000 rpm for 1 minute. Collection tube was discarded and column was placed in another collection tube.

8. **Second Wash**

500 µl of wash solution was added to the column and centrifuged for 3 minutes at 13500 rpm. Collection tube was discarded and the column was placed in a new 2 ml collection tube.

9. **Elution**

200 µl of the elution solution was added to the center of the column and centrifuged for 1 minute at 10,000 rpm.

The DNA sample collected was further subjected to qualitative, quantitative and integrity checks.

B.Qualitative, Quantitative and Integrity analysis of extracted DNA

Quantitative analysis:- Measurement at 260 nm by UV Spectroscopy

UV Spectrometer (Thermofischer) was used for qualitative and quantitative analysis of the DNA. Nucleic acids strongly absorbs at 260 nm whereas proteins absorb strongly at 280 nm. The general rules for determining the concentrations of nucleic acids at 260 nm are:

1. One optical Density (OD) unit of double-stranded DNA is 50 µg/ml.
2. One OD unit of single-stranded DNA is 33 µg/ml.
3. One OD unit of single-stranded RNA is 40 µg/ml.
4. Proteins absorb strongly at 280 nm where one OD unit is 1 mg/ml.

Procedure:-

Preparation of DNA sample:- In a sterile 1.5 ml centrifuge tube DNA sample was diluted 1 : 100 (dilution factor "100") with double distilled water and readings were taken at 260 and 280 nm on UV spectrophotometer.

DNA concentrations (µg/ml) = (OD_{260}) x (dilution factor) x 50 µg DNA/ml) / (1 OD_{260}unit)

Qualitative analysis:-The ratio of the absorbance at 260nm/ absorbance at 280 nm gave a measure of the quality of a DNA sample. Ratio of about 1.8 is considered as free of contamination.

The absorbance of a DNA sample at 280 nm gave an estimate of the protein contamination of the sample. As protein contamination increases, the ratio decreases.

Integrity analysis: - Agarose gel electrophoresis

1.5 g of agarose was weighed and mixed with 100 ml of tris borate EDTA (TBE) buffer in microwavable flask. The mixture was micro waved for 1-2 min till the agarose is completely dissolved. The agarose solution was cooled down for 5 min. 20 µl Ethidium bromide (EtBr) was added to obtain concentration of approximately 0.2 μg/ml. EtBr binds to the DNA and allows you to visualize the DNA under ultraviolet (UV) light. The gel was poured into a gel tray with the well comb in place. It was allowed to solidify at room temperature for 20-30 minutes, until it has completely solidified.

5 μl of extracted DNA with 6x loading dye. was transferred on agarose gel and run in TBE buffer at 100 volts for 45 minutes. The DNA with high integrity appeared as a clear distinct band. Loss of integrity resulted in smearing of DNA band. Figure 5.3 below illustrates the check gel electrophoresis for DNA integrity.

Preparation of 5X TBE buffer

S. No	Content	Weight/ Volume
1.	Tris Base (HIMEDIA)	54gm
2.	Boric Acid (Qualigens)	27.5gm
3.	0.5M EDTA (HIMEDIA)	20ml
Preparation of 0.5M EDTA		
1.	EDTA	93gm
Dissolved in 500ml of D/W (pH was adjusted to 8.0)		

Preparation of Loading Dye (6X)

1.	Bromophenol Blue (HIMEDIA	25mg
2.	Xylene Cyanol (HIMEDIA)	25mg
3.	Sucrose (HIMEDIA)	4gm
Final volume was made 10ml with D/W		

Preparation of Working Loading Dye

1.	6X loading dye	1ml

2.	D/W	5 ml

Preparation of 10% Ethidium Bromide (EtBr) Solution

10mg EtBr (HIMEDIA) was dissolved in 1ml of D/W.

Preparation of 1.5% Agarose gel

0.75gm of Agarose (HIMEDIA) was dissolved in 50 ml 1X TBE buffer.

Figure 5.3 DNA Gel electrophoresis (Integrity check)

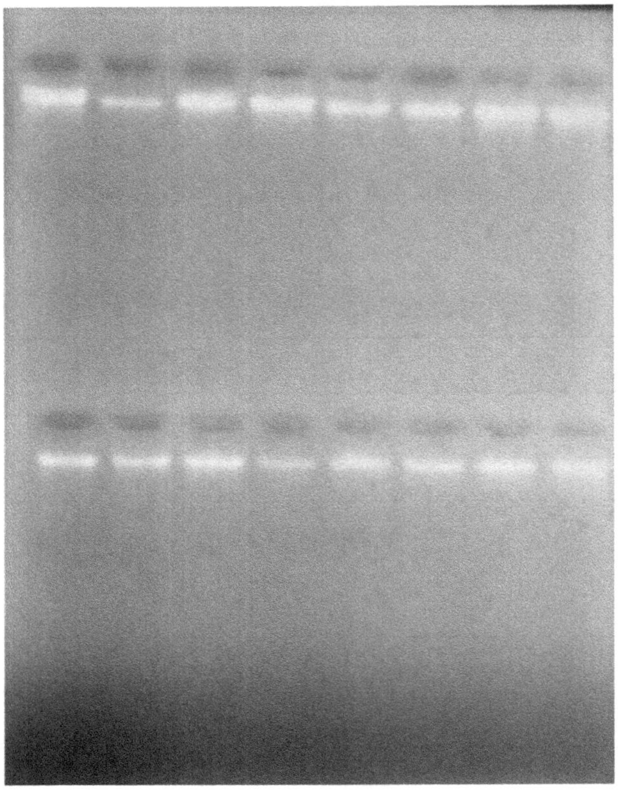

C. Polymerase Chain Reaction[333]

The sequence of the target DNA of eNOS gene (exon 7q35- 36) was verified on the NCBI site and the single nucleotide polymorphism (rs1799983) was verified.[335]

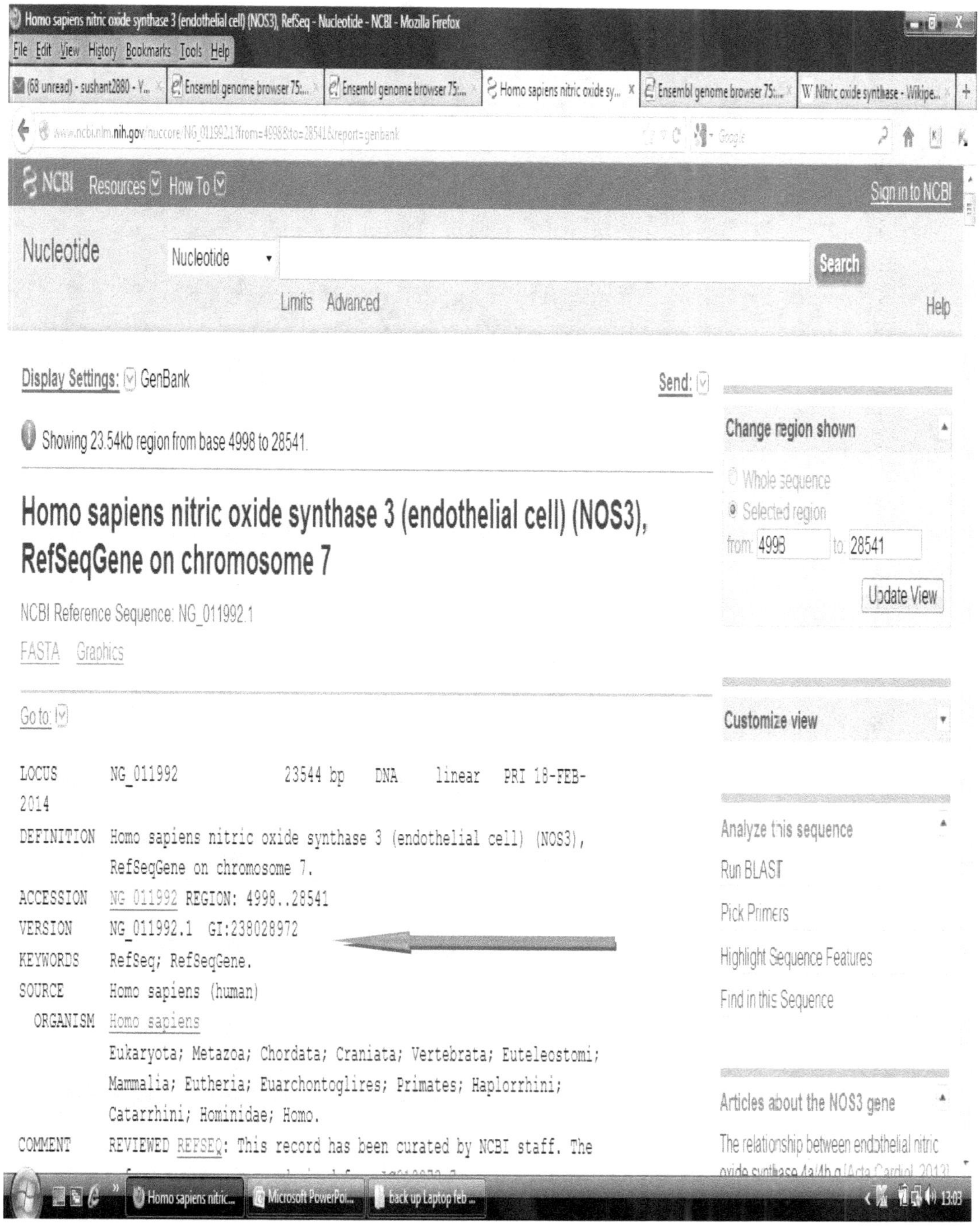

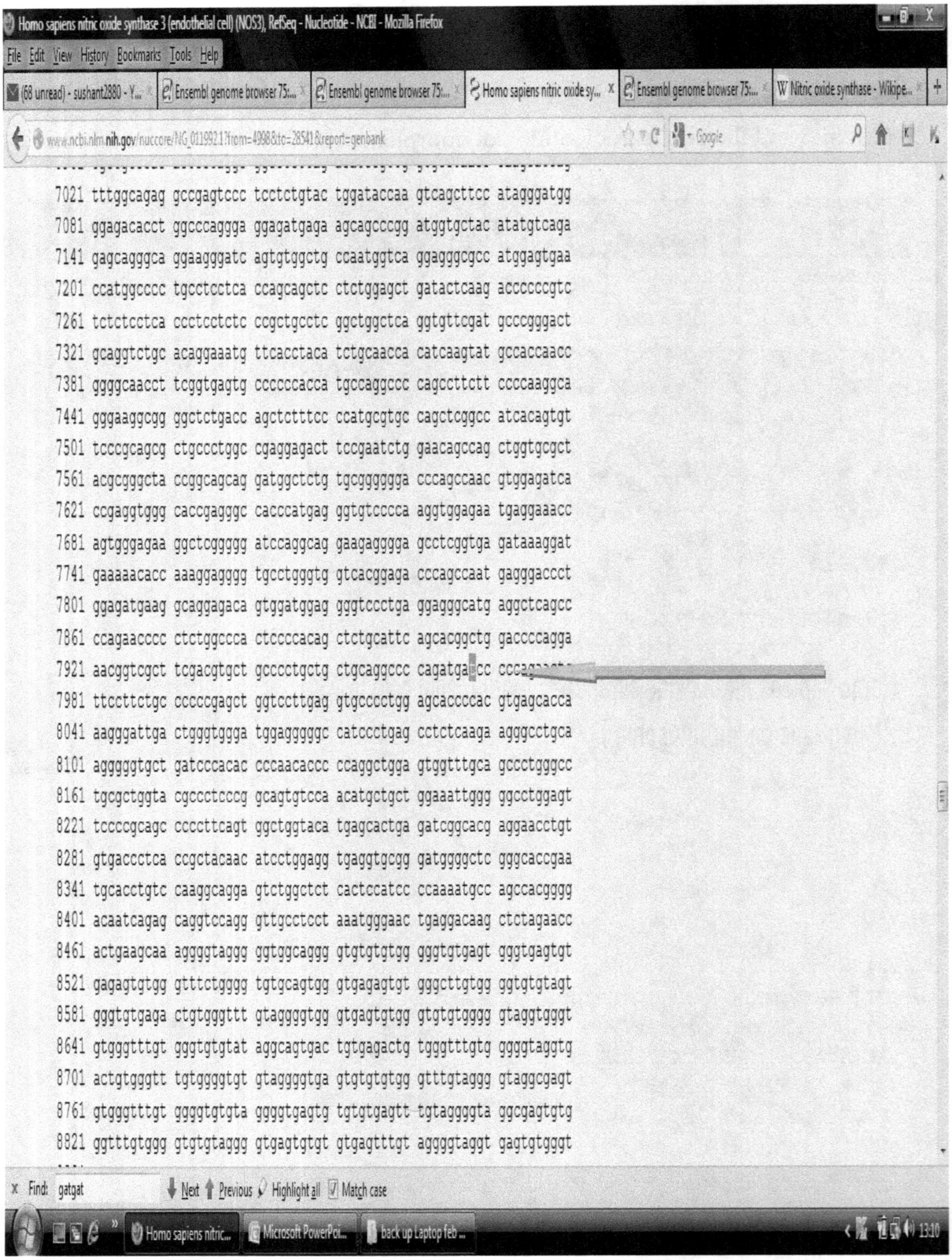
Homo sapiens nitric oxide synthase 3 (endothelial cell) (NOS3), RefSeq - Nucleotide - NCBI - Mozilla Firefox
File Edit View History Bookmarks Tools Help
(68 unread) - sushant2880 - Y...
Ensembl genome browser 75:...
Ensembl genome browser 75:...
Homo sapiens nitric oxide sy...
Ensembl genome browser 75:...
Nitric oxide synthase - Wikipe...
www.ncbi.nlm.nih.gov/nuccore/NG_011992.1?from=4998&to=28541&report=genbank
Google
7021 tttggcagag gccgagtccc tcctctgtac tggataccaa gtcagcttcc atagggatgg
7081 ggagacacct ggcccaggga ggagatgaga agcagcccgg atggtgctac atatgtcaga
7141 gagcagggca ggaagggatc agtgtggctg ccaatggtca ggagggcgcc atggagtgaa
7201 ccatggcccc tgcctcctca ccagcagctc ctctggagct gatactcaag acccccgtc
7261 tctctcctca ccctcctctc ccgctgcctc ggctggctca ggtgttcgat gcccgggact
7321 gcaggtctgc acaggaaatg ttcacctaca tctgcaacca catcaagtat gccaccaacc
7381 ggggcaacct tcggtgagtg ccccccacca tgccaggccc cagccttctt ccccaaggca
7441 gggaaggcgg ggctctgacc agctctttcc ccatgcgtgc cagctcggcc atcacagtgt
7501 tcccgcagcg ctgccctggc cgaggagact tccgaatctg gaacagccag ctggtgcgct
7561 acgcgggcta ccggcagcag gatggctctg tgcgggggga cccagccaac gtggagatca
7621 ccgaggtggg caccgagggc cacccatgag ggtgtcccca aggtggagaa tgaggaaacc
7681 agtgggagaa ggctcggggg atccaggcag gaagagggga gcctcggtga gataaaggat
7741 gaaaaacacc aaaggagggg tgcctgggtg gtcacggaga cccagccaat gagggaccct
7801 ggagatgaag gcaggagaca gtggatggag gggtccctga ggagggcatg aggctcagcc
7861 ccagaacccc ctctggccca ctccccacag ctctgcattc agcacggctg gaccccagga
7921 aacggtcgct tcgacgtgct gcccctgctg ctgcaggccc cagatga cc ccca
7981 ttccttctgc cccccgagct ggtccttgag gtgcccctgg agcaccccac gtgagcacca
8041 aagggattga ctgggtggga tggagggggc catccctgag cctctcaaga agggcctgca
8101 agggggtgct gatcccacac cccaacaccc ccaggctgga gtggtttgca gccctgggcc
8161 tgcgctggta cgccctcccg gcagtgtcca acatgctgct ggaaattggg ggcctggagt
8221 tccccgcagc ccccttcagt ggctggtaca tgagcactga gatcggcacg aggaacctgt
8281 gtgaccctca ccgctacaac atcctggagg tgaggtgcgg gatggggctc gggcaccgaa
8341 tgcacctgtc caaggcagga gtctggctct cactccatcc ccaaaatgcc agccacgggg
8401 acaatcagag caggtccagg gttgcctcct aaatgggaac tgaggacaag ctctagaacc
8461 actgaagcaa aggggtaggg ggtggcaggg gtgtgtgtgg gggtgtgagt gggtgagtgt
8521 gagagtgtgg gtttctgggg tgtgcagtgg gtgagagtgt gggcttgtgg ggtgtgtagt
8581 gggtgtgaga ctgtgggttt gtaggggtgg gtgagtgtgg gtgtgtgggg gtaggtgggt
8641 gtgggtttgt gggtgtgtat aggcagtgac tgtgagactg tgggtttgtg gggtaggtg
8701 actgtgggtt tgtggggtgt gtaggggtga gtgtgtgtgg gtttgtaggg gtaggcgagt
8761 gtgggtttgt ggggtgtgta ggggtgagtg tgtgtgagtt tgtaggggta ggcgagtgtg
8821 ggtttgtggg gtgtgtaggg gtgagtgtgt gtgagtttgt aggggtaggt gagtgtgggt
Find: gatgat
Next
Previous
Highlight all
Match case
Homo sapiens nitric...
Microsoft PowerPoi...
back up Laptop feb ...
13:10

The polymerase chain reaction (PCR) is a fast technique used to "amplify" - copy - small segments of DNA. It is a three-step process, referred to as a cycle, that is repeated a specified number of times. One PCR cycle consists of the following steps:

-Denaturation

-Annealing

-Extension

This process takes place in a thermal cycler, an instrument that automatically controls and alternates the temperatures for programmed periods of time for the appropriate number of PCR cycles (usually between 30 and 40 cycles).

To perform PCR amplification, all the components required for DNA replication are mixed as follows:

Template DNA:	The template carries the DNA segment or target to be amplified.
Primer:	A primer is a short, single-stranded piece of DNA that anneals (attaches) to its complementary sequence on the template. A pair of primers will bind to either side of the target DNA segment providing initiation sites for DNA synthesis. The two primers used in a PCR are often designated by the terms forward and reverse.
DNA Polymerase.	This is the enzyme used to synthesize new strands of DNA. DNA polymerase adds nucleotides onto the end of an annealed primer. The added bases are complementary to the template strand. DNA polymerase requires magnesium for activity which supplied to a PCR amplification in the form of magnesium chloride.
dNTPs.	dNTP stands for deoxynucleoside triphosphate which included equimolar mixture of dGTP. dATP,dCTP, dTTP. These are the four nucleotides used by DNA polymerase to extend an annealed primer.

Buffer.	A buffer is a solution that resists change in pH. It is used to ensure that the DNA polymerase is in an environment in which it will have maximum activity.

Components for PCR procedures

Sterile deionised water
10X PCR buffer
dNTP mix
Primer
Taq DNA polymerase
$MgCl_2$
Template DNA

Equipments for PCR procedures

Micropipettes
Thermocycler
Electrophoresis units
Power supply units
Photographic equipment

PCR Primer design and Tm calculation[336]

The PCR primers were designed using NCBI primer BLAST designing tool. The Tm calculation was done using Tm calculator for obtaining the GC concentration. All parameters were optimized for amplification of eNOS gene which included reagent concentrations, cycling temperatures and cycle number. The standadised conditions are mentioned in the protocol.

i) PCR Primers designing[29]

RESULTS

Dilution Resuspension

SEQUENCE:

5'- AAG GCA GGA GAC AGT GGA TGG ACC CCT CCA TCC CAC CCA GTC AAT C -3'

COMPLEMENT:

5'- GAT TGA CTG GGT GGG ATG GAG GGG TCC ATC CAC TGT CTC CTG CCT T -3'

LENGTH:	46
GC CONTENT:	58.7 %
MELT TEMP:	72.0 ℃
MOLECULAR WEIGHT:	14083.2 g/mole
EXTINCTION COEFFICIENT:	444100 L/(mole·cm)
nmole/OD_{260}:	2.25
µg/OD_{260}:	31.71

MELTING TEMPERATURE SETTINGS

TARGET TYPE:	DNA
OLIGO CONC	0.25 µM
Na^+ CONC	50 mM monovalent salt
Mg^{++} CONC	0 mM divalent salt
dNTPs CONC	0 mM nucleotide triphosphate

MELTING TEMPERATURE ASSUMPTIONS AND LIMITATIONS

Predictions are accurate for oligos from 8 to 60 bases in length, in neutral buffered solutions (pH 7 - 8) with monovalent cation (Na^+) concentrations from 1.2 M down to 1.5mM, divalent cation (Mg^{++}) concentrations from 600 mM down to 0.01 mM, and triphosphates (dNTPs) concentrations up to 120% of the divalent cation concentration.

Oligo concentration is assumed to be significantly larger (at least 6x) than concentration of the complementary target, which is true in majority of molecular biology experiments. If this is not a case, concentration of the target cannot be ignored and you should enter in the box,

Oligo Conc = [strand1] – [strand2]/2 when [strand1] ≥ [strand2]
Oligo Conc = ([strand1] + [strand2])/4 when [strand1] = [strand2]

Melting temperature accuracy and models: (Oligo/Template)

DNA/DNA: +/- 1.4°C (Allawi '97)
LNA/DNA: +/- 2.0°C (McTigue '04, Owczarzy, 2011)
RNA/DNA: +/- 2.7°C (Sugimoto '95)
DNA/RNA: +/- 2.7°C (Sugimoto '95)
RNA/RNA: +/- 1.3°C (Xia '98)
Divalent cation correction +/- 0.5°C (Owczarzy '08)
Triphosphate correction +/- 0.0°C (Owczarzy '08)
Monovalent cation correction: +/- 2.0°C (Owczarzy '04)

Consecutive LNA bases hybridized to a DNA template use a model from Owczarzy '11. In the absence of empirical data, LNA bases on an RNA template assume RNA values, and predictions are therefore less accurate.

Non-consecutive LNA bases hybridized to a DNA template use a model from McTigue '04. Consecutive LNA bases on a DNA template and any LNA bases on an RNA template assume RNA energetic parameters and predictions are therefore less accurate.

Effects of chemical modifications are neglected except when the modification contains a base, e.g., 5-Methyl dC, Internal Fluorescein dT. Energetic effects of these modifications are only approximated.

ii) Tm Calculator results of primer for PCR
Tm calculation results for these primers are as follows:

Primer	Tm (DegC)	Salt (mM)	Primer (µM)	%GC	Length
Primer #1: AAGGCAGGAGACAGTGGATGGA	64.02	50	.2	55	22
Primer #2: CCCCTCCATCCCACCCAGTCAATC	71.09	50	.2	63	24
Tm Delta = 7.07					

Melting Temperature (Tm) Calculation
Primer (6-50 bases):

CCCCTCCATCCCACCCAGTCAATC Calculate Tm

LENGTH 24
C+G% 62.5
Molecular weight: 7179.395

☑ Basic Tm
Degenerated nucleotides are allowed

Tm: **62.5 °C**

Melting Temperature (Tm) Calculation
Primer (6-50 bases):

AAGGCAGGAGACAGTGGATGGA Calculate Tm

LENGTH 22
C+G% 54.5
Molecular weight: 7003.305

☑ Basic Tm
Degenerated nucleotides are allowed

Tm: **56.7 °C**

PCR primers from (fermentas)	
forward primer	5' AAGGCAGGAGACAGTGGATGGA 3'
reverse primer	5' CCCCTCCATCCCACCCAGTCAATC 3'

iii) Procedure for Amplification of DNA segment by PCR

A 261 bp fragment of eNOS gene was amplified by PCR described by taking 50 ng of genomic DNA, 10 pmol/µl of primers. Super mix containing Taq polymerase, dNTPs, $MgCl_2$ was added to make a final volume of 30µl. PCR was cycled 35 times at 94°C for 45 sec, 66°C for 45 sec and 72°C for 45 sec, with initial 5 min denaturation at 94°C and final extension at 72°C for 6 min. The negative control was included in every experiment to monitor for contamination. The amplification of 261bp PCR product was verified on 1.5% agarose gel using 5µl of PCR product.

Components for PCR reaction

S.No.	Content	volume
1.	Super mix (Thermo fischer scientific)	15.0 µl
2.	Forward primer (eurofins)	1.0 µl
3.	Reverse primer (eurofins)	1.0 µl
4.	Genomic DNA	10.0 µl
5.	Nuclease-free water	3.0 µl
	Total reaction mix	30 µl

iv) PCR Cycle Conditions

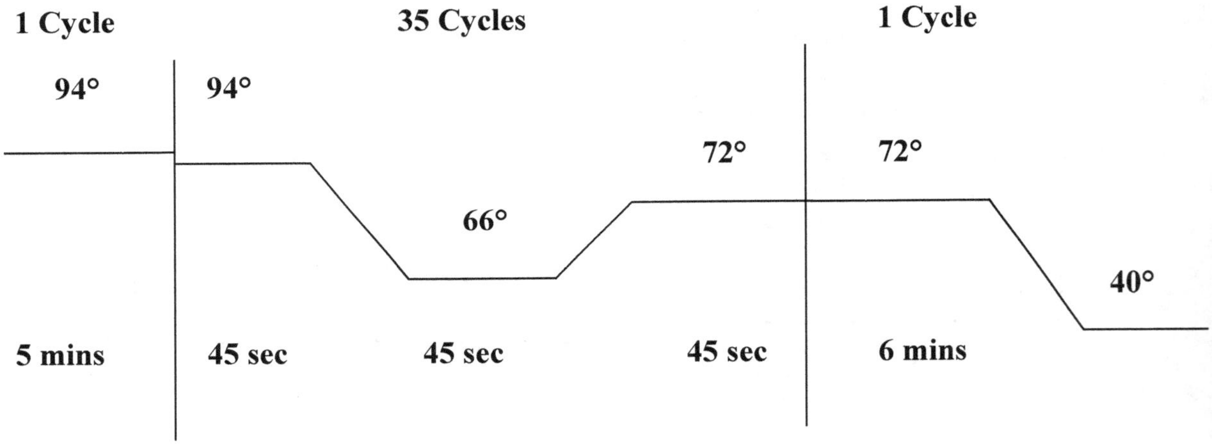

Figure 5.4 Agarose gel electrophoresis for PCR product

The post PCR amplified product was subjected to gel electrophoresis (1.5%) again to identify the size of the amplicon using appropriate ladder.

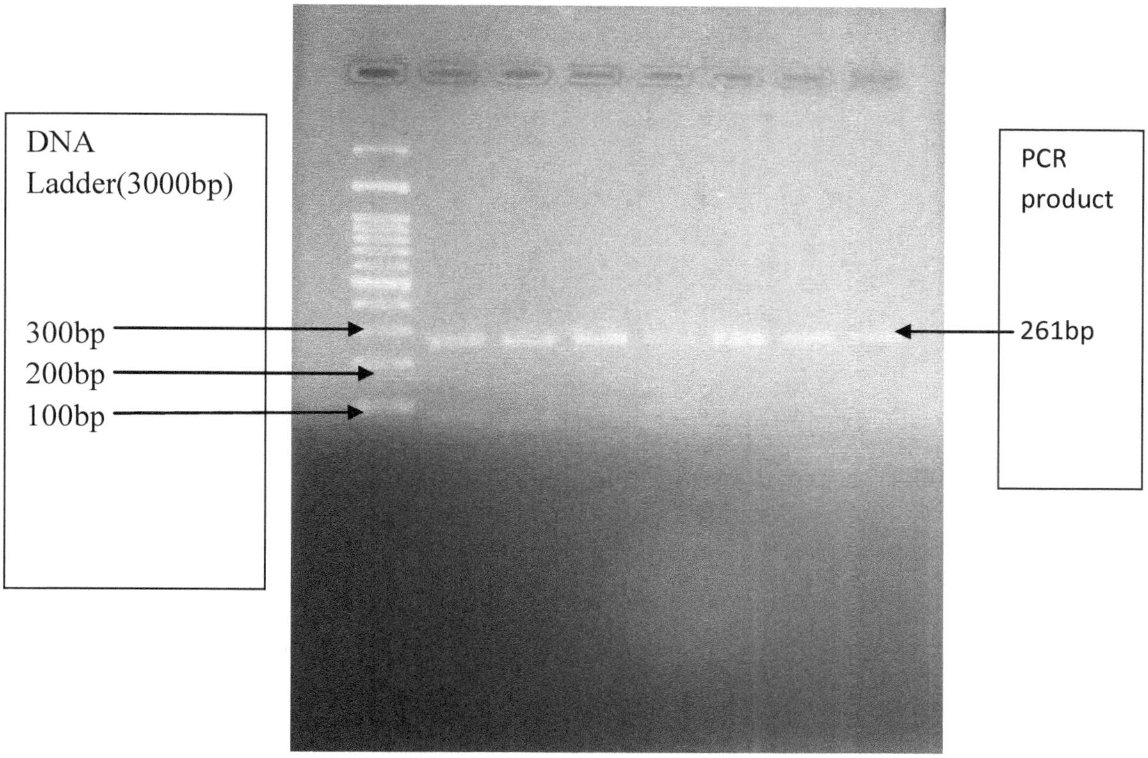

Figure 5.5 Conventional PCR machine (Thermo cycler peq start)

D. Restriction Fragment Length Polymorphism[26]

RFLP analysis is a methodology frequently used to identify genetic polymorphism after PCR amplification. The methodology in its simplest form involves the cutting of double stranded DNA molecules into smaller fragments using restriction enzymes. Restriction enzymes are enzymes that have been isolated from many different species of bacteria and are able to cut double stranded DNA by recognizing specific sequence (palindromic sequence usually between eight and six nucleotide in length) within DNA molecules. If the sequence of DNA molecule is known, then the size and number of fragment generated using a particular restriction enzyme may be predicted. By running the digested DNA product on a agarose gel, the theoretically predicted size and the number of fragments can be checked against the actual size and number of fragments obtained. If the observed sizes or number of fragments do not matched that predicted or expected, then fragment length polymorphisms are seen and variation (mutation) in the DNA sequence of that particular gene or amplimer exists. Using the RFLP technique it is possible to screen regions of DNA, which are known to undergo (frequent) mutations.

i)Determination of restriction enzyme for polymorphism by RFLP

Determination of nucleotide sequence was performed in following steps.

Step1:-The DNA sequence of amplified eNOS gene fragment was retrieved from NCBI Entrez nucleotide software.

Step2:- NEB cutter was used to determine the restriction enzyme for NG_011992.1 loci in the eNOS gene. The DNA sequence was entered in NEB cutter software and subjected to restriction digestion. The restriction enzyme, which cleaves the DNA fragment at the polymorphic loci, was tracked as shown below.

Custom digest of PCR product by Restriction enzyme (MboI)

Bands obtained after Custom digests of PCR product.[337]

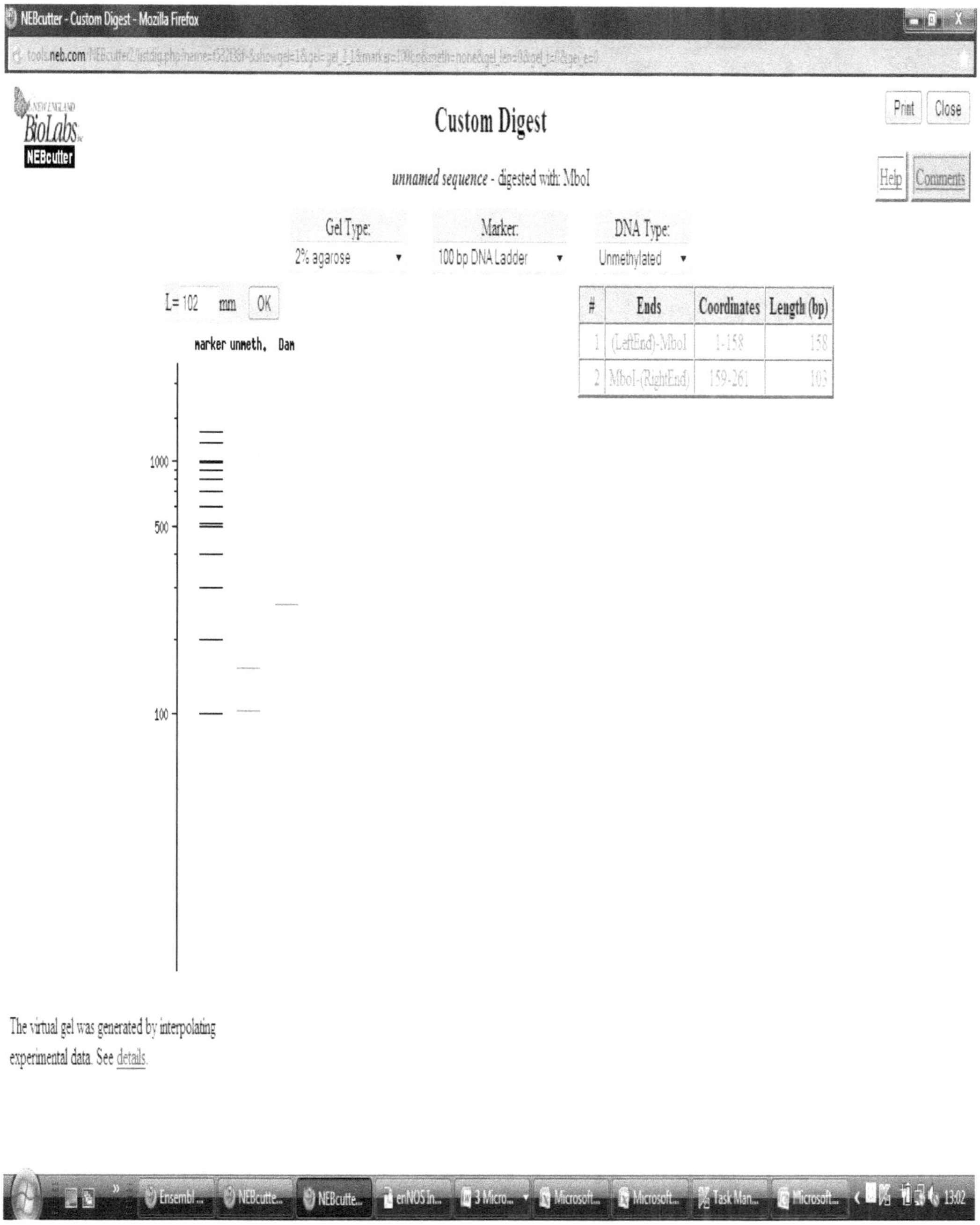

Digestion of PCR products

S.No	RFLP reaction mixture	Volume
1.	PCR reaction mixture	10ul
2.	Nuclease-free water	18 ul
3.	10X buffer tango	02 ul
4.	Mbo I (Thermoscietific; #ER0811)	01 ul (10 units/ul)

The above was mixed gently and incubated at 37° for 5 hours. The digested products was run on 1.5% agarose gel

Figure 5.6 Gel electrophoresis of RFLP

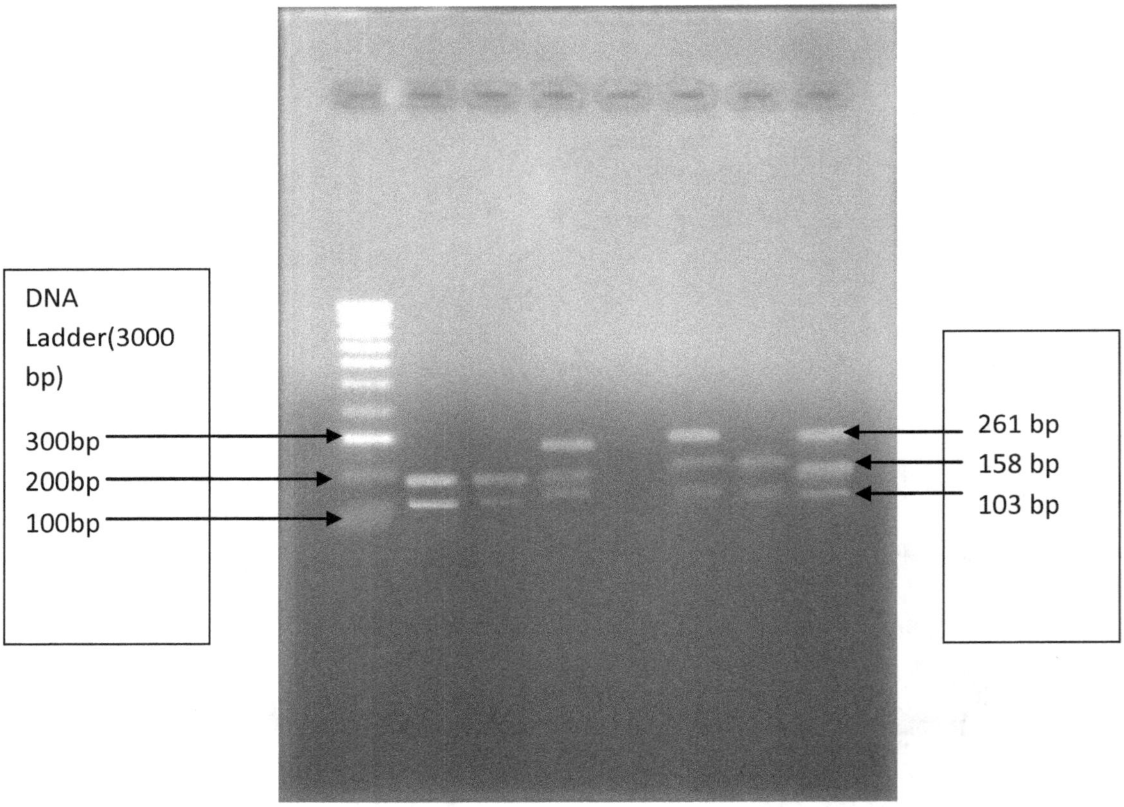

Figure 5.7 Gel doc (Lab India)

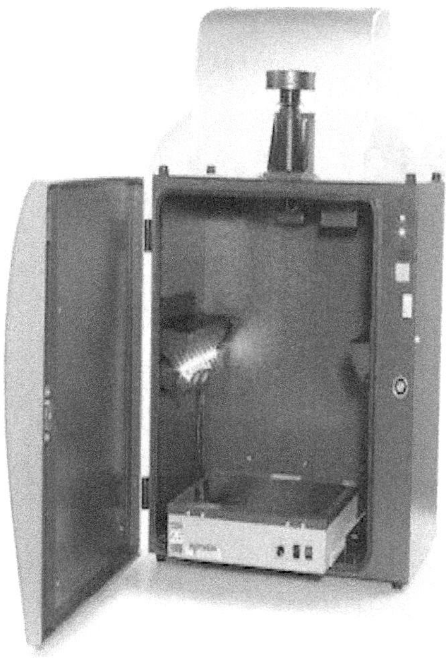

The above Gel doc system used to check the RFLP agarose electrophoresis pattern and get the photographs of stained gels.

Chapter 6

Result

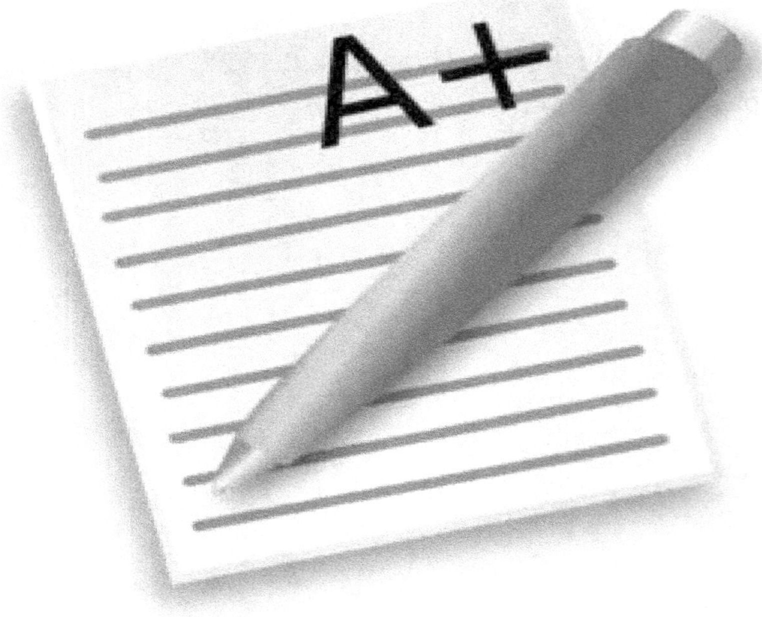

The present study consisted a study population of 230 which comprised of 117 males and 113 females in age group of 35 to 60 years. The population included healthy controls, patients of metabolic syndrome, type 2 diabetes and diabetes with complications. The entire population was divided in to four different groups. All the subjects of different groups were matched for age and gender.

The study mainly involved measurements of anthropometric parameters, estimation of biochemical analysis and genetic analysis. Biochemical analysis involved estimation of glycemic parameters, lipid parameters and endothelial parameters. Genetic analysis included isolation of genomic DNA and amplification of eNOS gene by polymerase chain reaction and study of single nucleotide polymorphism by restriction fragment polymorphism technique.

The results are presented in five sections

6.1 Anthropometric measurements
6.2 Glycemic parameters
6.3 Lipid parameters
6.4 Endithelial parameters
6.5 Genetic analysis

Section 6.1

Anthropometric parameters

- Anthropometric parameters involved measurement of weight, height, waist circumference, hip circumference, blood pressure.
- The waist to hip ratio (WHR) and body mass index (BMI) were calculated for all the controls and subjects.
- Table 6.1 shows descriptive anthropometric parameters of control and all the study groups.
- Graph 6.1 shows the pictorial comparisons of WHR, BMI and blood pressure of all the groups.
- Table 6.1a to 6.1f shows comparison of all anthropometric parameters among different groups along with *p* values. Data is expressed as Mean ± SD. p value < 0.05 is statistically significant.
- Pearson correlation coefficient was calculated by correlating HOMA IR with WHR and HOMA IR with BMI. The correlation analysis was done to study the association of WHR and BMI with HOMA IR i.e its effect on insulin resistance. The correlation graphs of different study groups are presented from figure 6.1a to 6.1f.
- The combined regression analysis was performed considering HOMA IR as dependent variable and WHR and BMI as independent variables in all study groups. The regression models are presented in table form from 6.1g, 6.1h and 6.1i.
- The Pearson correlation coefficients of HOMA IR with WHR and BMI of all study groups are presented in table 6.1j and 6.1k.

Table 6.1 Anthropometric parameters in different study Groups

Groups	BP (Systolic) mm/Hg	BP (Diastolic) mm/Hg	BMI in kg/m^2	WHR
Group I (Control) n=50	115 ± 6.98	76.7 ± 4.98	23.3 ± 1.39	0.802±0.03
Group II (MS) n=60	134 ± 2.86	96 ± 0.60	29.6 ± 0.53	0.99±0.01
Group III (T2DM) n=60	136.9 ±11.77	99.82± 11.66	30 ± 4.6	1.00±0.06
Group IV (T2DM with complications) n=60	145 ± 11.99	109± 13	30.1 ± 1.86	0.99 = 0.05

All data is expressed as Mean ±SD

Graph 6.1 Anthropometric parameters of all Groups

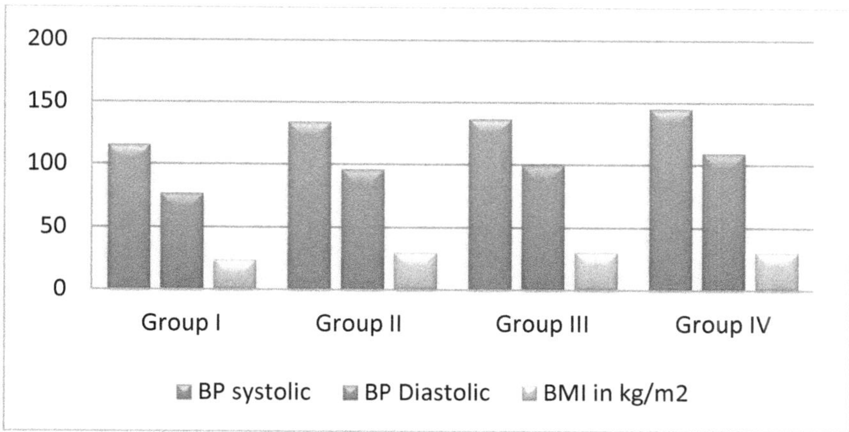

Comparison of Anthropometric parameters between control and study Groups using student t test

Table 6.1a Comparison of Anthropometric parameters of Group I with Group II

Groups	BP (Systolic) mm/Hg	BP (Diastolic) mm/Hg	BMI in kg/m^2	WHR
Group I (Control) n=50	115 ± 6.98	76.7 ± 4.98	23.3 ± 1.39	0.802 ±0.03
Group II (MS) n=60	134 ± 2.86	96 ± 0.60	29.6 ± 0.53	0.99±0.01
P value	0.0001*	0.0001*	0.0001*	0.0001*

All data is expressed as Mean ±SD
Student t test was applied to data to find statistical significance
*p<0.0001 is statistically significant

Table 6.1b Comparison of Anthropometric parameters of Group I and Group III

Groups	BP (Systolic) mm/Hg	BP (Diastolic) mm/Hg	BMI in kg/m^2	WHR
Group I (Control) n=50	115 ± 6.98	76.7 ± 4.98	23.3 ± 1.39	0.802 ±0.03
Group III (T2DM) n=60	136.9 ±11.77	99.82 ± 11.66	30 ± 4.6	1.00±0.06
P value	0.0001*	0.0001*	0.0001*	0.0001*

All data is expressed as Mean ±SD
Student t test was applied to data to find statistical significance
*p<0.0001 is statistically significant

Table 6.1c Comparison of Anthropometric parameters of Group I with Group IV

Groups	BP (Systolic) mm/Hg	BP (Diastolic) mm/Hg	BMI in kg/m^2	WHR
Group I (Control) n=50	115 ± 6.98	76.7 ± 4.98	23.3 ± 1.39	0.802 ±0.03
Group IV (T2DM with complications) n=60	145 ± 11.99	109± 13	30.1 ± 1.86	0.99 ± 0.05
P value	0.0001*	0.0001*	0.0001*	1.0

All data is expressed as Mean ±SD

Student t test was applied to data to find statistical significance

*p<0.0001 is statistically significant

Table 1d Comparison of Anthropometric parameters of Group II with Group III

Groups	BP (Systolic) mm/Hg	BP (Diastolic) mm/Hg	BMI in kg/m^2	WHR
Group II (MS) n=60	134 ± 2.86	96 ± 0.60	29.6 ± 0.53	0.99±0.01
Group III (T2DM) n=60	136.9 ±11.77	99.82 ± 11.66	30 ± 4.6	1.00±0.06
P value	0.09	0.01**	0.49	0.20

All data is expressed as Mean ±SD

Student t test was applied to data to find statistical significance

*p<0.0001, ** p<0.05 is statistically significant

Table 1e Comparison of Anthropometric parameters of Group II with Group IV

Groups	BP (Systolic) mm/Hg	BP (Diastolic) mm/Hg	BMI in kg/m^2	WHR
Group II (MS) n=60	134 ± 2.86	96 ± 0.60	29.6 ± 0.53	0.99±0.01
Group IV (T2DM with complications) n=60	145 ± 11.99	109± 13	30.1 ± 1.86	0.99 ± 0.05
P value	0.0001*	0.0001*	0.52	1.0

All data is expressed as Mean ±SD

Student t test was applied to data to find statistical significance

*$p<0.0001$ is statistically significant

Table 1f Comparison of Anthropometric parameters of Group III and Group IV

Groups	BP (Systolic) mm/Hg	BP (Diastolic) mm/Hg	BMI in kg/m^2	WHR
Group III (T2DM) n=60	136.9 ±11.77	99.82 ± 11.66	30 ± 4.6	1.00±0.06
Group IV (T2DM with complications) n=60	145 ± 11.99	109± 13	30.1 ± 1.86	0.99 ± 0.05
P value	0.0001*	0.0001*	0.43	0.32

All data is expressed as Mean ±SD

Student t test was applied to data to find statistical significance

*$p<0.0001$ is statistically significant

Figure 6.1a Correlation between HOMA IR and WHR in Group II

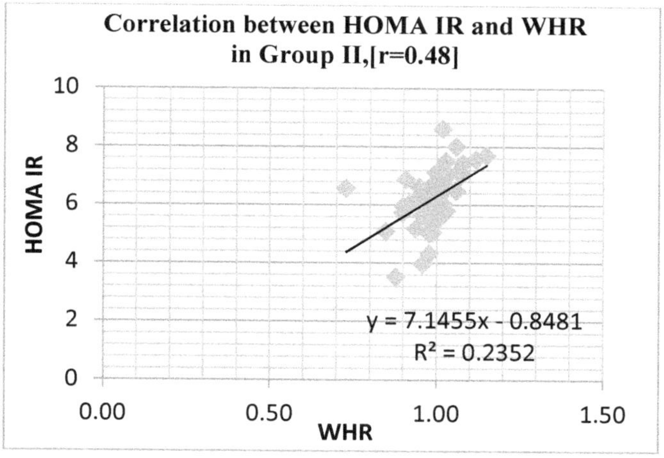

Figure 6.1b Correlation between HOMA IR and WHR in Group III

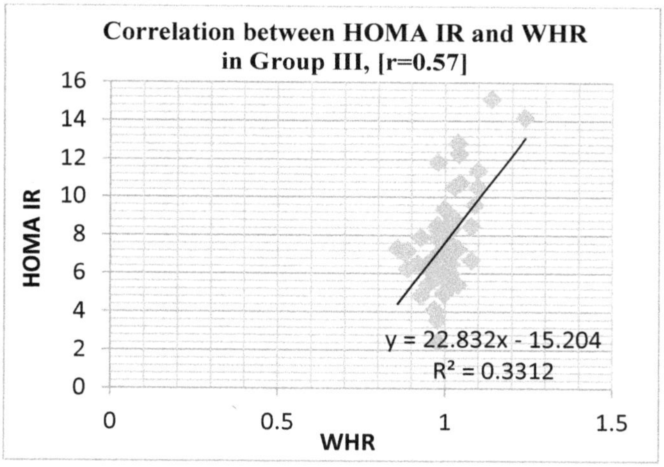

Figure 6.1c Correlation between HOMA IR and WHR in Group IV

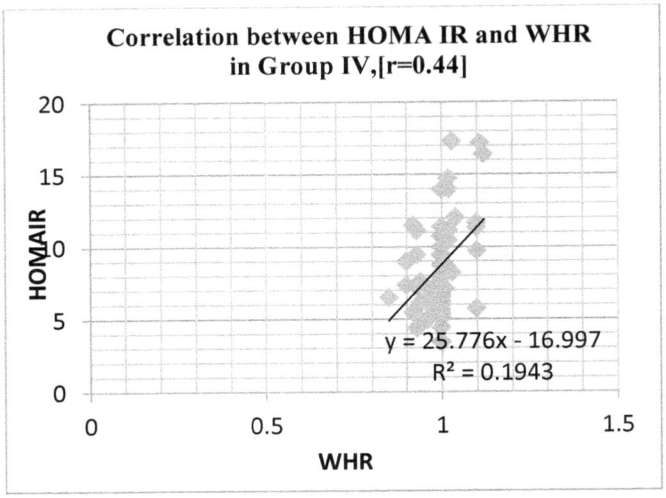

Figure 6.1d Correlation between HOMA IR and BMI in Group II

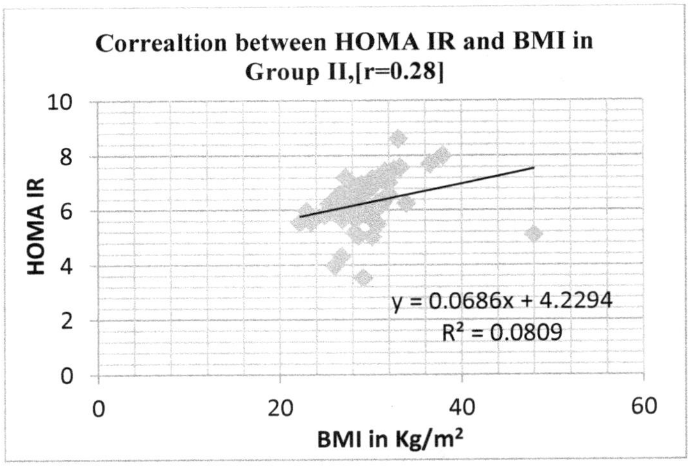

Figure 6.1e Correlation between HOMA IR and BMI in Group III

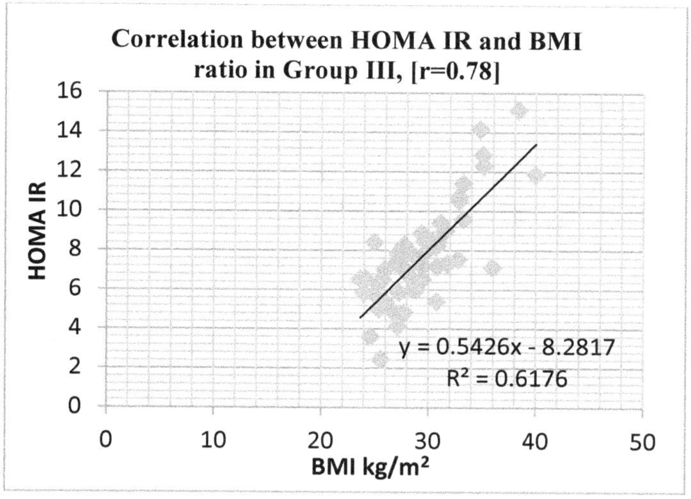

Figure 6.1f Correlation between HOMA IR and BMI in Group IV

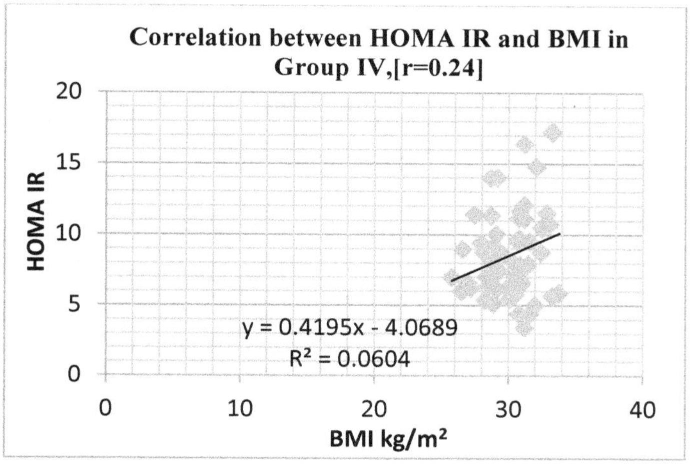

Table 6.1g Linear regression model betwwen HOMA IR and WHR and BMI in Group II

Linear Regression Model					
Dependent Variable: HOMA IR					
Variable	**Reg. Coefficient**	**SE**	**P-value**	**Correlation Coefficient**	**R^2**
WHR	6.18	1.52	0.001	0.54	0.29
BMI	0.07	0.03	0.03		

Table 6.1h Linear regression model between HOMA IR and WHR and BMI in Group III

Linear Regression Model					
Dependent Variable: HOMA IR					
Variable	**Reg. Coefficient**	**SE**	**P-value**	**Correlation Coefficient**	**R^2**
WHR	0.30	4.64	0.04	0.71	0.51
BMI	7.98	0.06	0.001		

Table 6.1i Linear regression model between HOMA IR and WHR and BMI in Group IV

Linear Regression Model					
Dependent Variable: HOMA IR					
Variable	**Reg. Coefficient**	**SE**	**P-value**	**Correlation Coefficient**	**R^2**
WHR	20.89	6.19	0.001	0.47	0.22
BMI	0.31	0.20	0.13		

Table 6.1j Correlation coefficients of HOMA IR and WHR in all Groups

Study Groups	r	R^2	p
Group I	**0.11**	**0.01**	**0.43**
Group II	**0.48**	**0.23**	**0.0001**
Group III	**0.57**	**0.33**	**0.0001**
Group IV	**0.44**	**0.19**	**0.0006**

Table 6.1k Correlation coefficients of HOMA IR and BMI in all Groups

Study Groups	r	R^2	p
Group I	**0.08**	**0.007**	**0.55**
Group II	**0.28**	**0.08**	**0.02**
Group III	**0.78**	**0.617**	**0.0001**
Group IV	**0.24**	**0.060**	**0.06**

Section 6.2

Glycemic Parameters

- Glycemic parameters included estimation of fasting and post prandial plasma glucose, HbA1c and fasting plasma insulin.
- Insulin resistance was calculated as HOMA IR in all the study groups.
- Table 6.2 shows descriptive analysis of glycemic parameters in all groups. All the study groups (II,III and IV) showed elevated fasting and postprandial plasma sugar, increased percentage of HbA1c, hyperinsulinemia and insulin resistance. Graph 6.2a depicts fasting and postprandial plasma glucose levels in all groups. Graph 6.2b showed comparison between HbA1c, plasma insulin and HOMA IR in all the groups.
- Intergroup comparison of glycemic parameters is depicted in tables 6.2a to 6.2f along with *p* values. Data is expressed and Mean ± SD. p value < 0.05 is statistically significant.

Table 6.2 Glycemic parameters in different study Groups

Groups	FBG mg/dl	PP mg/dl	HbA1c %	Fasting plasma insulin µIU/ml	HOMA IR
Group I (Control) n=50	88.1 ±5.06	123.2 ± 4.91	4.34 ± 0.33	10.2 ± 3.2	2.23 ± 0.69
Group II (MS) n=60	113 ± 1.19	149 ± 31	5.8 ± 0.5	22 ± 0.11	6.14 ± 0.09
Group III (T2DM) n=60	207±59.3	288 ± 94	7.52± 1.28	16.12±3.93	7.71 ±2.43
Group IV (T2DM with complications) n=60	276 ± 108	365±132	7.9 ± 0.9	13.0 ± 3.5	8.57±3.17

All data is expressed as Mean ± SD

Graph 6.2a Glycemic parameters (FPG and PPPG) of all Groups

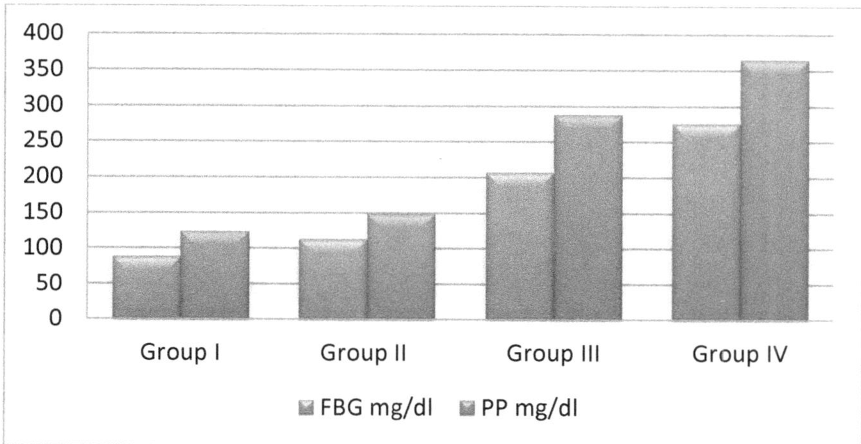

Graph 6.2b Glycemic parameters (HbA1C, Plasma Insulin and HOMA IR) of all Groups

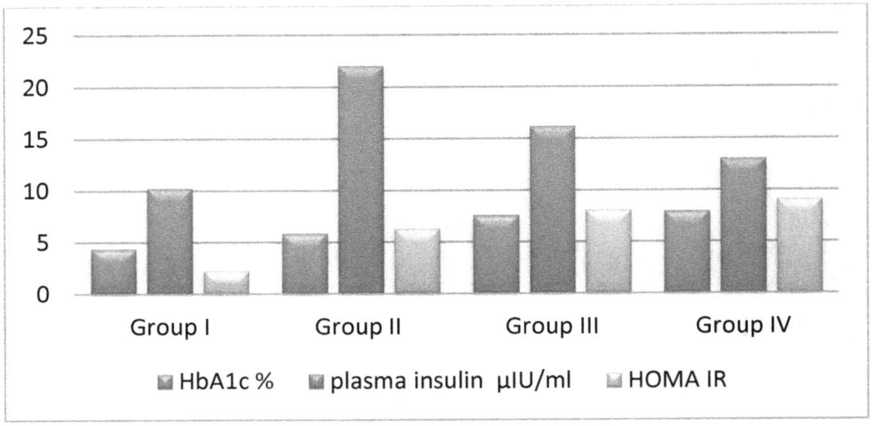

Comparison of Glycemic parameters between control and study Groups using student t test

Table 6.2a Comparison of Glycemic parameters of Group I with Group II

Groups	FPG mg/dl	PPPG mg/dl	HbA1c	Fasting plasma insulin µIU/ml	HOMA IR
Group I (Control) n=50	88.1 ± 5.06	123.2 ± 4.91	4.34 ± 0.33	10.2 ± 3.2	2.23 ± 0.69
Group II (MS) n=60	113 ± 1.19	149 ± 31	5.8 ± 0.5	22 ± 0.11	6.14 ± 0.09
P value	0.0001*	0.0001*	00001*	0.0001*	0.0001*

All data is expressed as Mean ± SD
Student t test was applied to data to find statistical significance
*p<0.0001 is statistically significant

Table 6.2b Comparison of Glycemic parameters of Group I with Group III

Groups	FPG mg/dl	PPPG mg/dl	HbA1c	Fasting plasma insulin µIU/ml	HOMA IR
Group I (Control) n=50	88.1 ± 5.06	123.2 ± 4.91	4.34 ± 0.33	10.2 ± 3.2	2.23 ± 0.69
Group III(T2DM) n=60	207±59.3	288 ± 94	7.52± 1.28	16.12±3.93	7.71 ±2.43
P value	0.0001*	0.0001*	0.0001*	0.0001*	0.0001*

All data is expressed as Mean ± SD
Student t test was applied to data to find statistical significance
*p<0.0001 is statistically significant

Table 6.2c Comparison of Glycemic parameters of Group I with Group IV

Groups	FPG mg/dl	PPPG mg/dl	HbA1c	Fasting Plasma insulin µIU/ml	HOMA IR
Group I (Control) n=50	88.1 ± 5.06	123.2 ± 4.91	4.34 ± 0.33	10.2 ± 3.2	2.23 ± 0.69
Group IV (T2DM with complications) n=60	276 ± 108	365±132	7.9 ± 0.9	13± 3.5	8.57±3.17
P value	0.0001*	0.0001*	0.0001*	0.0001*	0.0001*

All data is expressed as Mean ± SD
Student t test was applied to data to find statistical significance
*p<0.0001 is statistically significant

Table 6.2d Comparison of Glycemic parameters of Group II with Group III

Groups	FPG mg/dl	PPPG mg/dl	HbA1c	Fasting plasma insulin µIU/ml	HOMA IR
Group II (MS) n=60	113 ± 1.19	149 ± 31	5.8 ± 0.5	22 ± 0.11	6.14 ± 0.09
Group III (T2DM) n=60	207±59.3	288 ± 94	7.52± 1.28	16.12±3.93	7.71 ±2.43
P value	0.0001*	0.0001*	0.0001*	0.0001*	0.0001*

All data is expressed as Mean ± SD
Student t test was applied to data to find statistical significance
*p<0.0001 is statistically significant

Table 6.2e Comparison of Glycemic parameters of Group II with Group IV

Groups	FPG mg/dl	PPPG mg/dl	HbA1c	Fasting plasma insulin µIU/ml	HOMA IR
Group II (MS) n=60	113± 1.19	149 ± 31	5.8 ± 0.5	22 ± 0.11	6.14 ± 0.09
Group IV (T2DM with complications) n=60	276 ± 108	365±132	7.9 ± 0.9	13± 3.5	8.57±3.17
P value	0.0001*	0.0001*	0.0001*	0.0001*	0.0001*

All data is expressed as Mean ± SD
Student t test was applied to data to find statistical significance
*p<0.0001 is statistically significant

Table 6.2f Comparison of Glycemic parameters of Group III and Group IV

Groups	FPG mg/dl	PP mg/dl	HbA1c	Fasting plasma insulin µIU/ml	HOMA IR
Group III (T2DM) n=60	207±59.3	288 ± 94	7.52± 1.28	16.12±3.93	7.71 ±2.43
Group IV (T2DM with complications) n=60	276 ± 108	365±132	7.9 ± 0.9	13± 3.5	8.57±3.17
P value	0.0001*	0.0001*	0. 07	0.001**	0.10

All data is expressed as Mean ± SD
Student t test was applied
*p<0.01 is statistically significant

Section 6.3

Lipid Parameters

- Study of lipid parameters included analysis of total cholesterol (TC), triglycerides (TG), high density lipoprotein cholesterol (HDL-C) and apolipoprotein B levels in all the controls and subjects.
- The values of low density total cholesterol (LDL) and very low density cholesterol were calculated using Friewald's formula.
- Lipid parameters are shown in table 6.3 including total cholesterol, TG, LDL cholesterol, VLDL and Apo B. All the parameters except HDL were increased in all the study groups in comparison with controls. Pictorial comparison of lipid profile is shown in Graph 6.3a and 6.3b.
- The comparison of various lipid parameters among different study groups are represented in table 6.3a to 6.3f. Data is expressed as Mean ± SD.

  p value < 0.05 is statistically significant.
- The correlation studies of HOMA IR with TG/HDL-C and HOMA IR with Apo B were done in all study groups and are shown graphically in figures 6.3a, 6.3b, 6.3c and 6.3d, 6.3e, 6.3f.
- A multivariate analysis was performed between HOMA IR, a dependent variable and TG-HDL-C ratio as well as Apo B in all the groups. The regression models are depicted in the form of tables 6.3g. 6.3h and 6.3i repectively
- The correlation coefficients for all the groups are shown in table 6.3j and 6.3k

Table 6.3 Lipid parameters in different study Groups

Groups	Cholesterol mg/dl	TG mg/dl	HDL mg/dl	VLDL mg/dl	LDL mg/dl	Apo B mg/dl
Group I (Control) n=50	146.3±10.59	110± 17.2	44.5 ±3.6	21.59± 2.72	83.92± 14.2	82.6± 15.34
Group II (MS) n=60	174 ±15	189 ±7.5	40±3.7	36±2.1	103±22.2	109 ± 23
Group III (T2DM) n=60	205±33.3	216±73	38±5.5	46±32	128±34	158± 59
Group IV (T2DMwith complications) n=50	220 ±59	228±79	37±5.5	49±17	136±53.6	203.1± 76.6

All data is expressed as Mean ± SD

Graph 6.3a Lipid parameters (TC, TG, HDL) of all Groups

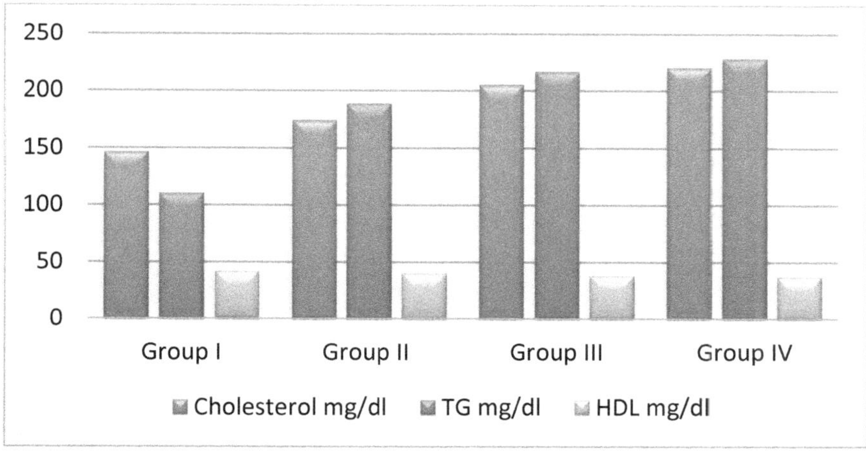

Graph 6.3b Lipid parameters (VLDL, LDL, Apo B) of all Groups

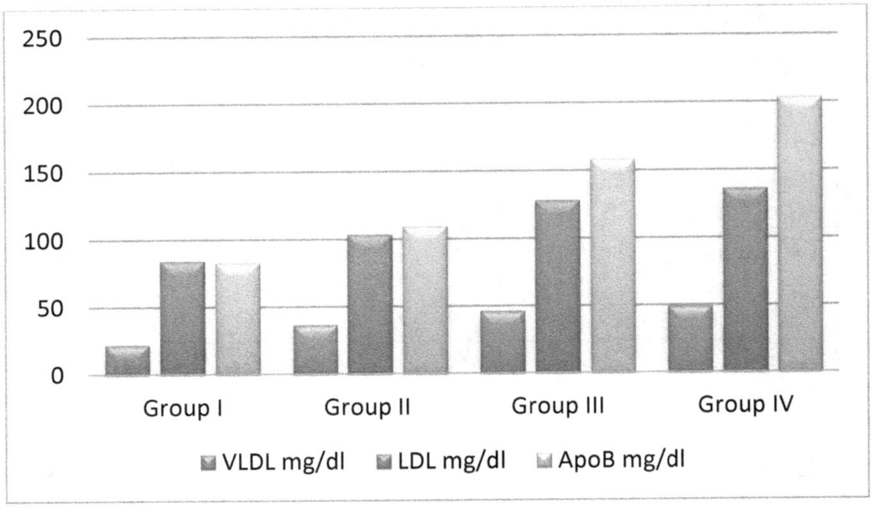

Comparison of Lipid Parameters between control and study Groups using student t test

Table 6.3a Comparison of Lipid parameters of Group I with Group II

Groups	Cholesterol mg/dl	TG mg/dl	HDL mg/dl	VLDL mg/dl	LDL mg/dl	Apo B mg/dl
Group I (Control) n=50	146.3±10.59	110± 17.2	44.5 ±3.6	21.59± 2.72	83.92± 14.2	82.62± 15.34
Group II (MS) n=60	174 ±15	189 ±7.5	40±3.7	36±2.1	103± 22.2	109± 23
P value	0.0001*	0.0001*	0.0001*	0.0001*	0.0001*	0.0001*

All data is expressed as Mean ± SD
Student t test was applied to data to find statistical significance
*p<0.0001 is statistically significant

Table 6.3b Comparison of Lipid parameters of Group I with Group III

Groups	Cholesterol mg/dl	TG mg/dl	HDL mg/dl	VLDL mg/dl	LDL mg/dl	Apo B mg/dl
Group I (Control) n=50	146.3±10.59	110±17.2	44.5 ±3.6	21.59± 2.72	83.92± 14.2	92±13
Group III (T2DM) n=60	205±33.3	216±73	38±5.5	46±32	128±34	158± 59
P value	0.0001*	0.0001*	0.0004*	0.0001*	0.0001*	0.0001*

All data is expressed as Mean ± SD
Student t test was applied to data to find statistical significance
*p<0.0001 is statistically significant

Table 6.3c Comparison of Lipid parameters of Group I with Group IV

Groups	Cholesterol mg/dl	TG mg/dl	HDL mg/dl	VLDL mg/dl	LDL mg/dl	Apo B mg/dl
Group I (Control) n=50	146.3±10.59	110± 17.2	44.5 ±3.6	21.59± 2.72	83.92± 14.2	92±13
Group IV (T2DM with complications) n=60	220 ±59	228± 79	37±5.5	49±17	136± 53.6	203.11± 76.6
P value	0.0001*	0.0001*	0.0001	0.0001*	0.0001*	0.0001*

All data is expressed as Mean ± SD

Student t test was applied to data to find statistical significance

*p<0.0001 is statistically significant

Table 6.3d Comparison of Lipid parameters of Group II with Group III

Groups	Cholesterol mg/dl	TG mg/dl	HDL mg/dl	VLDL mg/dl	LDL mg/dl	Apo B mg/dl
Group II (MS) n=60	174 ±15	189 ±7.5	40±3.7	36±2.1	103±22.2	109±23
Group III (T2DM) n=60	205±33.3	216±73	38±5.5	46±32	128±34	158± 59
P value	0.0001*	0.0001*	0.02**	0.0001*	0.0001*	0.0001*

All data is expressed as Mean ± SD

Student t test was applied to data to find statistical significance

*p<0.0001, **p<0.05 is statistically significant.

Table 6.3e Comparison of Lipid parameters of Group II with Group IV

Groups	Cholesterol mg/dl	TG mg/dl	HDL mg/dl	VLDL mg/dl	LDL mg/dl	Apo B mg/dl
Group II (MS) n=60	174 ±15	189 ±7.5	40±3.7	36±2.1	103±22.2	109 ±23
Group IV (T2DM with complications) n=60	220 ±59	228±79	37±5.5	49±17	136±53.6	203.1±76.6
P value	0.0001*	0.0001*	0.0001	0.0001*	0.0001*	0.0001*

All data is expressed as Mean ± SD

Student t test was applied to data to find statistical significance

*p<0.0001 is statistically significant

Table 6.3f Comparison of Lipid parameters of Group III and Group IV

Groups	Cholesterol mg/dl	TG mg/dl	HDL mg/dl	VLDL mg/dl	LDL mg/dl	Apo B mg/dl
Group III (T2DM) n=60	205±33.3	216±73	38±5.5	46=32	128±34	158± 59
Group IV (T2DM with complications) n=60	220 ±59	228±79	37±5.5	49±17	136±53.6	203.1±76.6
P value	0.08	0.39	0.31	0.53	0.33	0.01**

All data is expressed as Mean ± SD

Student t test was applied to data to find statistical significance

**p<0.05,*p<0.0001 is statistically significant

Figure 6.3a Correlation between HOMA IR and TG/HDL-C ratio in Group II

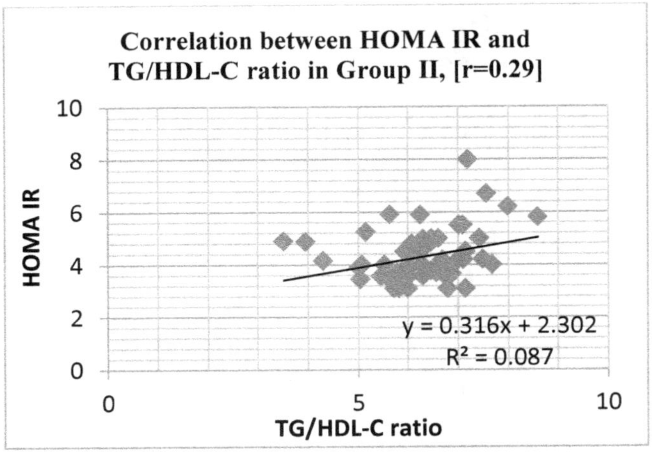

Figure 6.3b Correlation between HOMA IR and TG/HDL-C ratio in Group III

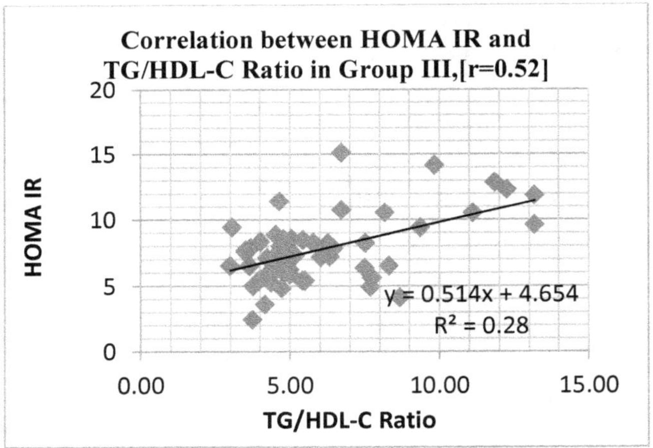

Figure 6.3c Correlation between HOMA IR and TG/HDL-C ratio in Group IV

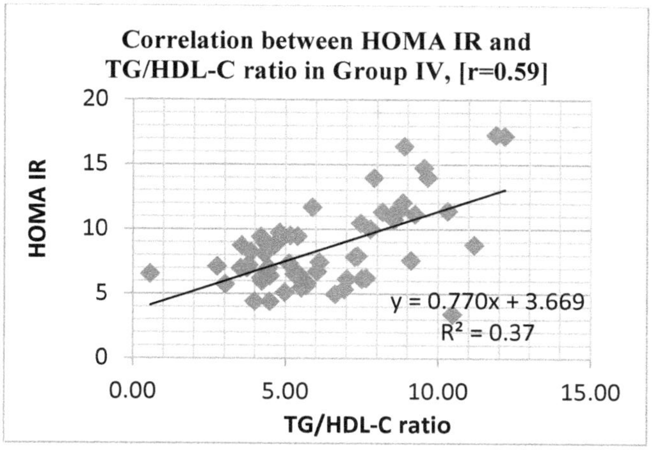

Figure 6.3d Correlation between HOMA IR and serum Apo B in Group II

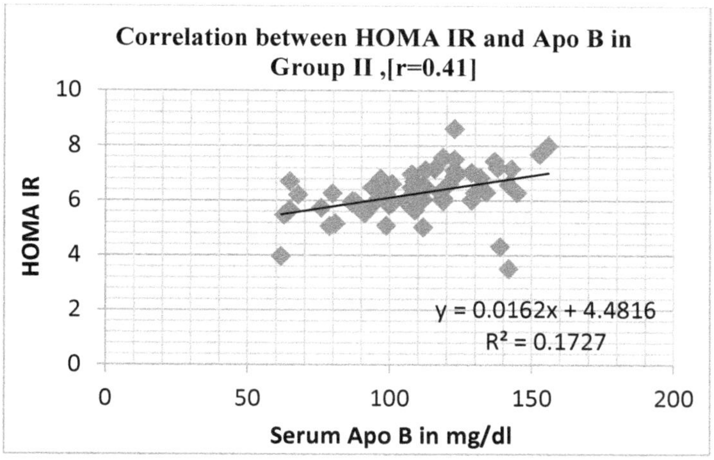

Figure 6.3e Correlation between HOMA IR and serum Apo B in Group III

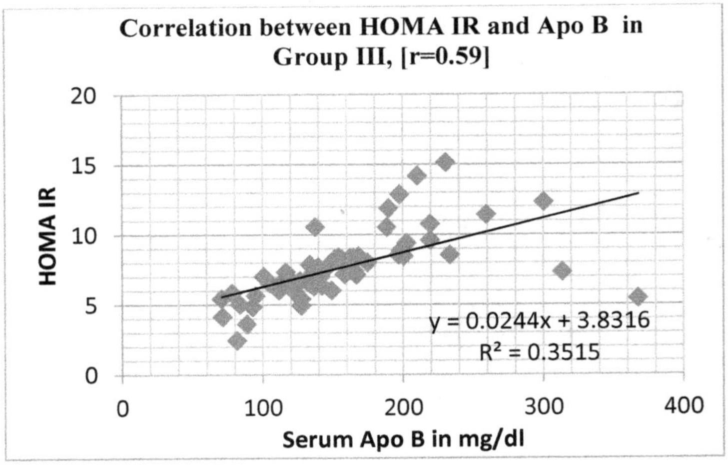

Figure 6.3f Correlation between HOMA IR and serum Apo B in Group IV

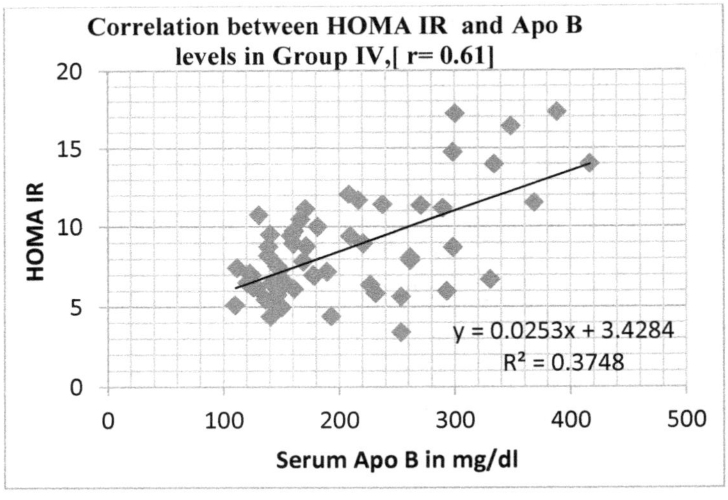

Table 6.3g Linear regression model between HOMA IR and TG/HDL-C ratio and Apo B in Group II

Linear Regression Model					
Dependent Variable: HOMA IR					
Variable	**Reg. Coefficient**	**SE**	**P-value**	**Correlation Coefficient**	R^2
TG/HDL-C	0.180	0.114	0.120	0.455	0.207
APO B	0.014	0.005	0.005		

Table 6.3h Linear regression model between HOMA IR and TG/HDL-C ratio and Apo B in Group III

Linear Regression Model					
Dependent Variable: HOMA IR					
Variable	**Reg. Coefficient**	**SE**	**P-value**	**Correlation Coefficient**	R^2
TG/HDL-C	0.369	0.103	0.0007	0.67	0.45
APO B	0.018	0.004	0.0001		

Table 6.3i Linear regression model betwwen HOMA IR and TG/HDL-C ratio and Apo B in Group IV

Linear Regression Model					
Dependent Variable: HOMA IR					
Variable	**Reg. Coefficient**	**SE**	**P-value**	**Correlation Coefficient**	**R^2**
TG/HDL-C	0.01	0.004	0.02	0.64	0.41
APO B	0.64	0.14	0.001		

Table 6.3j Correlation coefficients of HOMA IR and TG/HDL-C in all Groups

Study Groups	r	R^2	p
Group I	0.05	0.00	0.74
Group II	0.29	0.087	0.02
Group III	0.52	0.28	0.0001
Group IV	0.59	0.37	0.0001

Table 6.3k Correlation coefficients of HOMA IR and Apo B in all Groups

Study Groups	r	R^2	p
Group I	0.29	0.08	0.06
Group II	0.41	0.17	0.0001
Group III	0.59	0.35	0.0001
Group IV	0.61	0.37	0.0001

Section 6.4

Endothelial Parameters

- Serum nitric oxide and urinary microalbuminuria were the two endothelial markers studied in the present study in all the Groups.
- Table 6.4 presents endothelial parameters. Marked reduction in serum nitric oxide levels were seen in all study groups in comparison with control. The urinary microalbumin levels were statistically significant in diabetic group and diabetic with complications group.
- Comparison of serum nitric oxide levels and urine microalbumin levels in all groups is shown in graph 4a and 4b resp. Data is expressed as Mean ± SD. p value < 0.05 is statistically significant.
- The univariate regression analysis of HOMA IR with serum nitric oxide and urinary microalbumin was carried out in all the study groups and is presented in figure 6.4a, 6.4b, 6.4c and 6.4d, 6.4e, 6.4f.
- The multivariate analysis was done considering HOMA IR as a dependent variable and serum nitric oxide and urinary microalbumin as independent variables. The regression analysis models are depicted in tables 6.4g, 6.4h and 6.4i.
- The correlation coefficients for all the groups are shown in table 6.4j and 6.4k.

Table 6.4 Endothelial parameters in different study Groups

Groups	Nitric oxide µmol/l	Microalbumin mg/dl
Group I (Control) n=50	0.29 ± 0.052	5.02 ± 1.86
Group II (MS) n=60	0.27±.05	5.52±3.26
Group III (T2DM) n=60	0.16±0.051	61.09±27.56
Group IV (T2DM with complications) n=50	0.14±0.04	139.5 ±88.7

All data is expressed as Mean ± SD

Graph 6.4a Endothelial parameters (Serum Nitric oxide) of all Groups

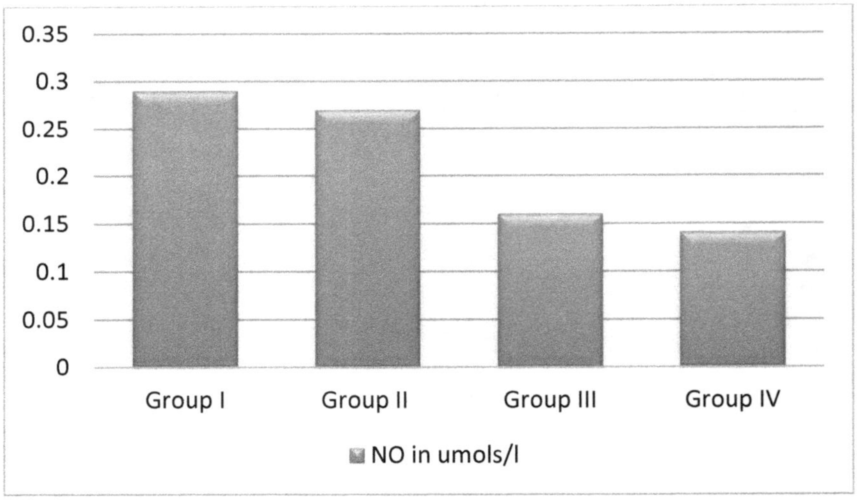

Graph 6.4b Endothelial parameters (Urinary Microalbumin) of all Groups

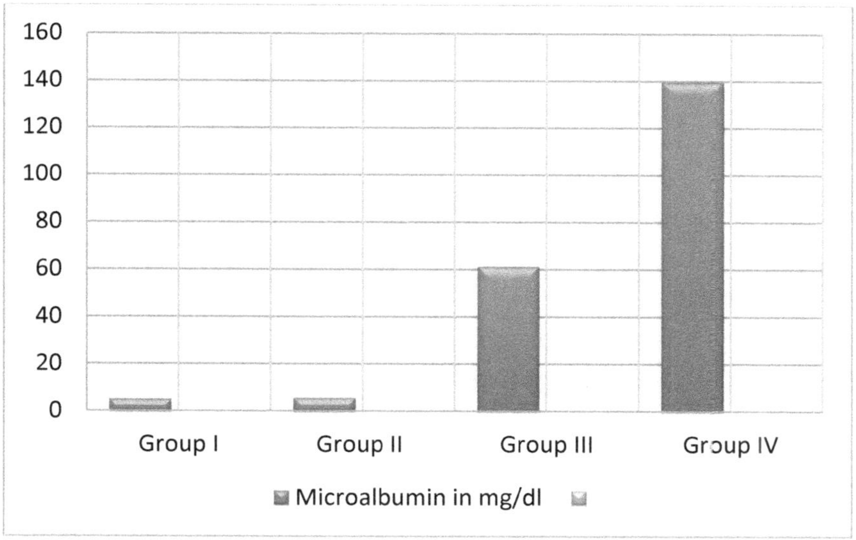

Comparison of endothelial parameters between control and study groups using student t test

Table 6.4a Comparison of Endothelial parameters of Group I and Group II

Groups	Serum Nitric oxide µmol/l	Urinary Microalbumin mg/dl
Group I (Control) n=50	0.29 ± 0.05	5.02 ± 1.86
Group II (MS) n=60	0.27±.05	5.5± 3.26
P value	0.03**	0.07

All data is expressed as Mean ± SD
Student t test was applied to data to find statistical significance
*P<0.0001, **P<0.05 is statistically significant

Table 6.4b Comparison of Endothelial parameters of Group I and Group III

Groups	Serum Nitric oxide µmol/l	Urinary Microalbumin mg/dl
Group I (Control) n=50	0.29 ± 0.05	5.05± 1.86
Group III (T2DM) n=60	0.16±0.05	61.14±27.56
P value	0.0001*	0.0001*

All data is expressed as Mean ± SD
Student t test was applied to data to find statistical significance
*P<0.0001, **P<0.05 is statistically significant

Table 6.4c Comparison of Endothelial parameters of Group I and Group IV

Groups	Serum Nitric oxide µmol/l	Urinary Microalbumin mg/dl
Group I (Control) n=50	0.29 ± 0.05	5.05± 1.86
Group IV (T2DM with complications) n=60	0.14±0.04	139±88.72
P value	0.01*	0.0001*

All data is expressed as Mean ± SD

Student t test was applied to data to find statistical significance

*P<0.0001, **P<0.05 is statistically significant

Table 6.4d Comparison of Endothelial parameters of Group II and Group III

Groups	Serum Nitric oxide µmol/l	Urinary Microalbumin mg/dl
Group II (MS) n=60	0.27±0.05	5.52±3.26
Group III (T2DM) n=60	0.16±0.054	61.14±27.56
P value	0.0001*	0.0001*

All data is expressed as Mean ±SD

Student t test was applied to data to find statistical significance

*P<0.0001, **P<0.05 is statistically significant

Table 6.4e Comparison of Endothelial parameters of Group II and Group IV

Groups	Serum Nitric oxide µmol/l	Urinary Microalbumin mg/dl
Group II (MS) n=60	0.27±0.05	5.52±3.26
Group IV (T2DM with complications) n=60	0.14±0.04	139±88.72
P value	0.0001*	0.0001*

All data is expressed as Mean ±SD

Student t test was applied to data to find statistical significance

*P<0.0001, **P<0.05 is statistically significant

Table 6.4f Comparison of Endothelial parameters of Group III and Group IV

Groups	Serum Nitric oxide µmol/l	Urinary Microalbumin mg/dl
Group III (T2DM) n=60	0.16±0.05	61.09±27.56
Group IV (T2DM with complications) n=60	0.14±0.04	139±88.7
P value	0.017*	0.0001*

All data is expressed as Mean ±SD

Student t test was applied to data to find statistical significance

*P<0.0001, **P<0.05 is statistically significant

Figure 6.4a Correlation between HOMA IR and serum Nitric Oxide in Group II

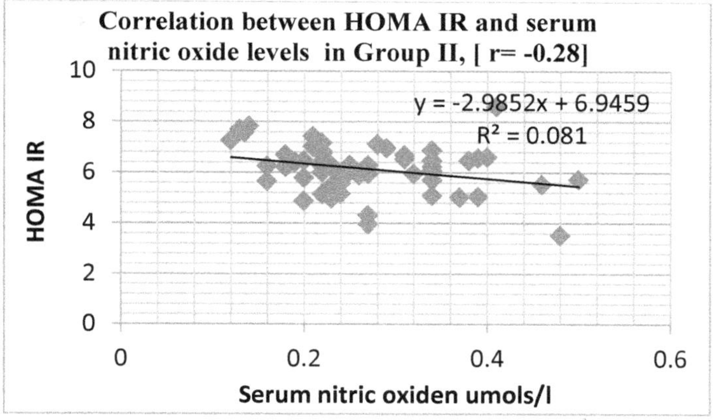

Figure 6.4b Correlation between HOMA IR and serum Nitric Oxide in Group III

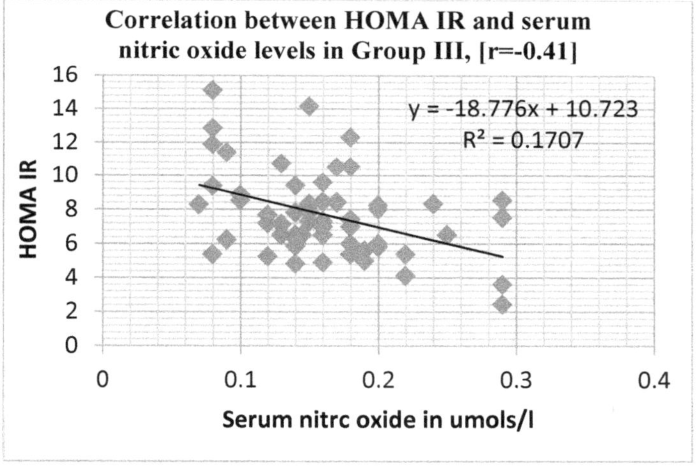

Figure 6.4c Correlation between HOMA IR and serum Nitric Oxide in Group IV

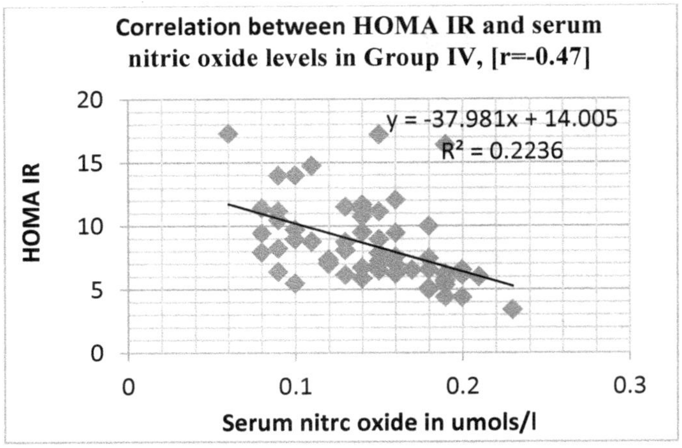

Figure 6.4d Correlation between HOMA IR and urinary microalbumin levels in Group II

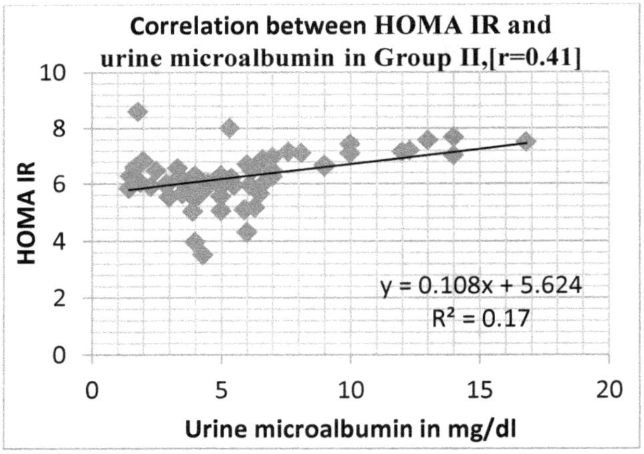

Figure 6.4e Correlation between HOMA IR and urinary microalbumin levels in Group III

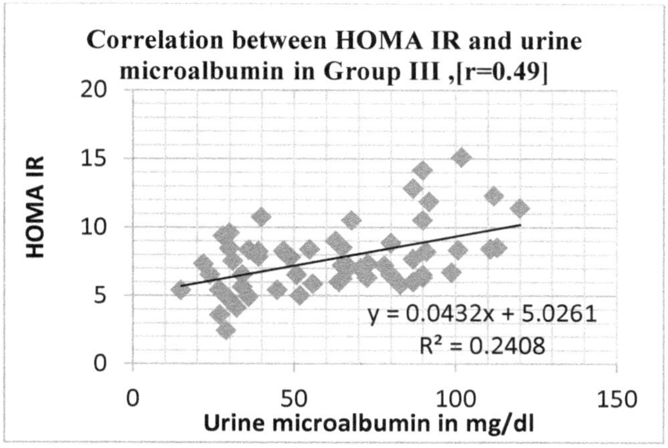

Figure 6.4f Correlation between HOMA IR and urinary microalbumin levels in Group IV

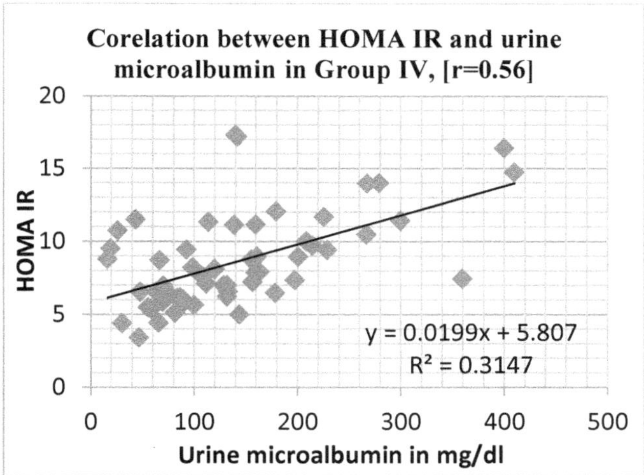

Table 6.4g Linear regression model between HOMA IR and serum nitric oxide and urinary microalbumin Group II

Linear Regression Model					
Dependent Variable: HOMA IR					
Variable	**Reg. Coefficient**	**SE**	**P-value**	**Correlation Coefficient**	**R^2**
Urinary microalbumin	7.28	0.37	0.01	0.42	0.18
Serum Nitric oxide	-0.08	0.03	0.046		

Table 6.4h Linear regression model between HOMA IR and serum nitric oxide and urinary microalbumin Group III

Linear Regression Model					
Dependent Variable: HOMA IR					
Variable	**Reg. Coefficient**	**SE**	**P-value**	**Correlation Coefficient**	**R^2**
Urinary microalbumin	7.81	1.206	0.006	0.58	0.34
Serum Nitric oxide	0.036	0.01	0.007		

Table 6.4i Linear regression model between HOMA IR and serum nitric oxide and urinary microalbumin Group IV

Linear Regression Model					
Dependent Variable: HOMA IR					
Variable	**Reg. Coefficient**	**SE**	**P-value**	**Correlation Coefficient**	**R^2**
Urinary microalbumin	0.01	0.003	0.0001	0.63	0.40
Serum Nitric oxide	-26.27	8.80	0.004		

Table 6.4j showing correlation coefficient of HOMA IR and serum nitric oxide in all Groups

Study Groups	r	R^2	p
Group I	0.089	0.008	0.53
Group II	-0.28	0.08	0.02
Group III	-0.41	0.17	0.0001
Group IV	-0.47	0.22	0.0001

Table 6.4k showing correlation coefficient of HOMA IR and urinary microalbumin in all Groups

Study Groups	r	R^2	p
Group I	0.01	0.00	0.92
Group II	0.41	0.17	0.0001
Group III	0.49	0.24	0.0001
Group IV	0.56	0.31	0.0001

Section 6.5

Genetic Analysis

- Genetic analysis was done to study single nucleotide polymorphism of eNOS gene in metabolic syndrome, type 2 diabetes patients with and without complication and its association with phenotype.
- The table 6.5a depicted the frequency of GG, GT & TT genotype in control and study groups. The frequency of (GG+GT) genotypes was more in the study group when compared with control group and there is association with metabolic syndrome.
- The allelic frequency of the genotypes is presented in table 6.5b. The risk of occurrence of disease is shown in table 6.5c.
- The genotype:phenotype association of within group analysis and inter group analysis is presented in table 6.5d, 6.5e.
- Between group comparison of genotype and phenotype is shown in table 6.5f.
- Figure 6.5a shows the gel electrophoresis pattern of restriction fragment length polymorphism of eNOS gene.
- Graph 6.5a, 6.5b, 6.5c and 6.5d shows the genotype frequency distribution of eNOS gene polymorphism in group I, II, III and IV respectively.

Table 6.5a Genotype frequency of SNP at G894T

Groups	GG	GT	TT
Control group I (n=50)	42(84)	08(16)	-
Metabolic syndrome Group II (n=60)	39(65)	16(27)	05(8)
Type 2 diabetes Group III (n=60)	32 (53)	18(30)	10(17)
Type 2 diabetes with complications Group IV(n=60)	25(42)	21(35)	14(23)

Table 6.5b Allelic frequency of SNP at G894T

Allele	Group I	Group II	Group III	Group IV
G Allele	90 %	78%	68%	59%
T Allele	10 %	22%	32%	41%s

Figure 6.5a Gel electrophoresis of RFLP

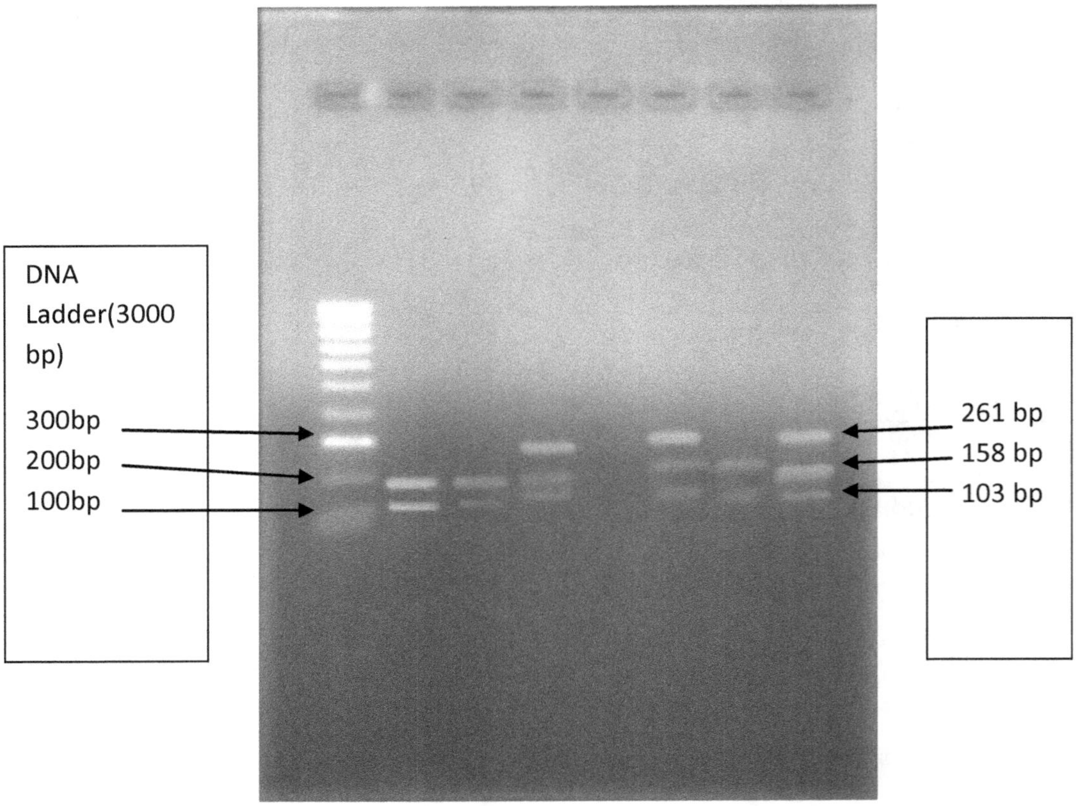

Table 6.5c Association of eNOS gene polymorphism with disease

Groups	Genotypes			χ^2 Test	Fischer Exact Test	
	GG	GT	TT		Odds Ratio 95% CI	P value
Control group I(n=50)	84%	16%	-	-		
Metabolic syndrome Group II (n=60)	65%	27%	8%	5.07	2.82 (1.12 - 7.11)	<0.05
Type 2 diabetes Group III (n=60)	53%	30%	17%	7.35	4.74 (1.9-11.8)	<0.001
Type 2 diabetes with complications Group IV (n=60)	42%	35%	23%	11.65	7.06 (2.84-17.56)	<0.000

Table 6.5d Genotype Association study

Parameters	Genotype	Control (Group I)	MS (Group II)	T2DM (Group III)	T2DM with complications (Group IV)
SerumNitric Oxide μmol/l	G/G	0.33±0.04	0.29±0.09	0.16±0.05	0.13±0.03
	G/T+T/T	0.32±0.04	0.22±0.06	0.15±0.04	0.14±0.04
Urinary Micro-albuminuria mg/dl	G/G	4.93±1.48	5.81±3.77	58.3±29.7	136.6±65.1
	G/T+T/T	5.47±2.58	5.0±2.1	62.5±26.6	141.4±104.2
HOMA IR	G/G	2.2±0.66	5.97±0.95	7.25±1.95	8.79±2.96
	G/T+T/T	2.2±2.32	6.47±0.73	8.1±2.89	8.41±3.38

Genotype specific Mean ±SD for each group

Table 6.5e Within Group pair wise Comparison

Parameters	Control (Group I)	MS (Group II)	T2DM (Group III)	T2DM with complications (Group IV)
	GG vs GT+TT	GG vs GT+TT	GG vs GT+TT	GG vs GT+TT
SerumNitric Oxide, µmol/l	P<0.89	P<0.02	P<0.99	P<0.35
Urinary Microalbuminuria, mg/dl	P<0.39	P<0.35	P<0.55	P<0.86
HOMA-IR	P<0.63	P<0.037	P<0.15	P<0.65

Table 6.5f Between group pair wise comparison

Parameters	Genotype	Control vs MS P value	Metabolic syndrome vs T2DM P value	T2DM vs T2DM with comp
SerumNitric Oxide, µmol/l	GG	P<1.0	P<0.001	P<0.4
	GT+TT	P<0.001	P<0.007	P<1.0
Urinary Microalbuminuria, mg/dl	GG	P<0.82	P<0.001	P<0.001
	GT+TT	P<1.0	P<0.001	P<0.001
HOMA IR	GG	P<0.001	P<0.018	P<0.25
	GT+TT	P<0.001	P<0.009	P<1.0
Based on estimated marginal means				
a. Adjustment for multiple comparisons: Bonferroni.				
*. The mean difference is significant at the .05 level.				

Graph 6.5a Frequency distribution of eNOS gene polymorphism in Group I

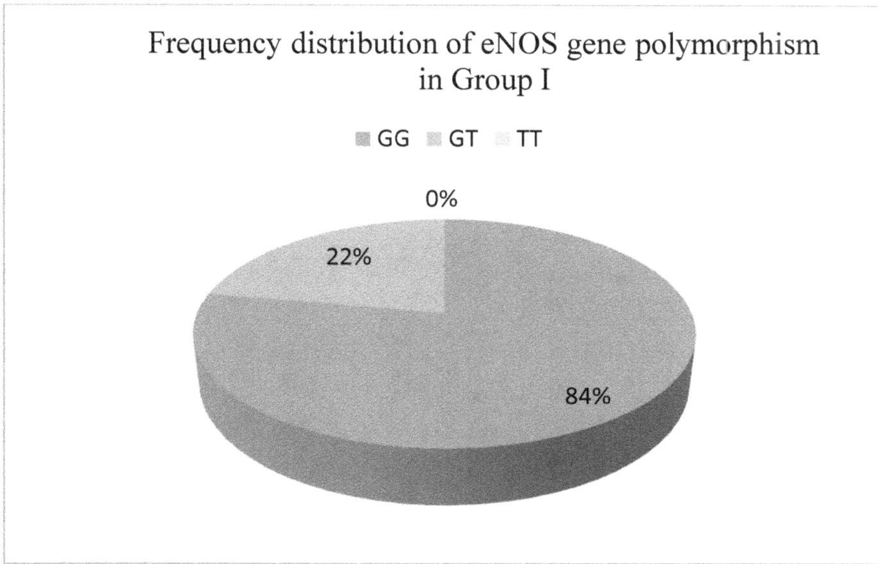

Graph 6.5b Frequency distribution of eNOS gene polymorphism in Group II

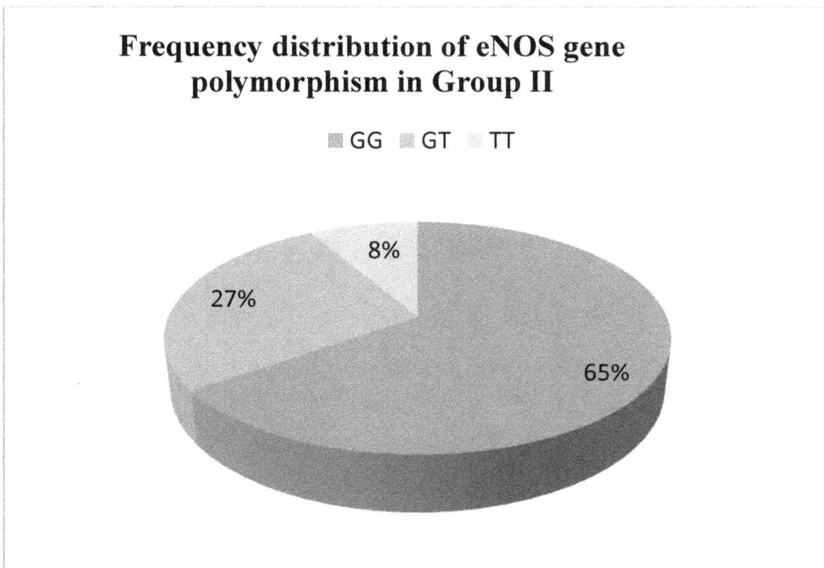

Graph 6.5c Frequency distribution of eNOS gene polymorphism in Group III

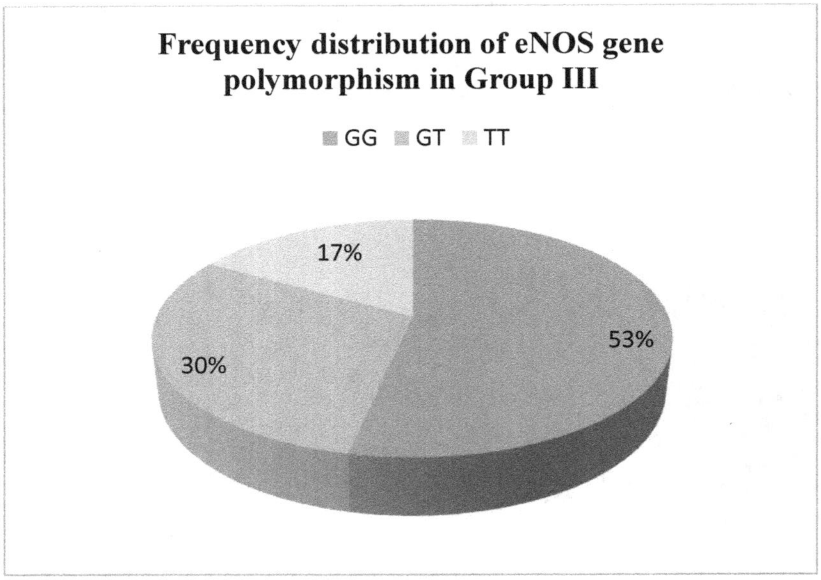

Graph 6.5d Frequency distribution of eNOS gene polymorphism in Group IV

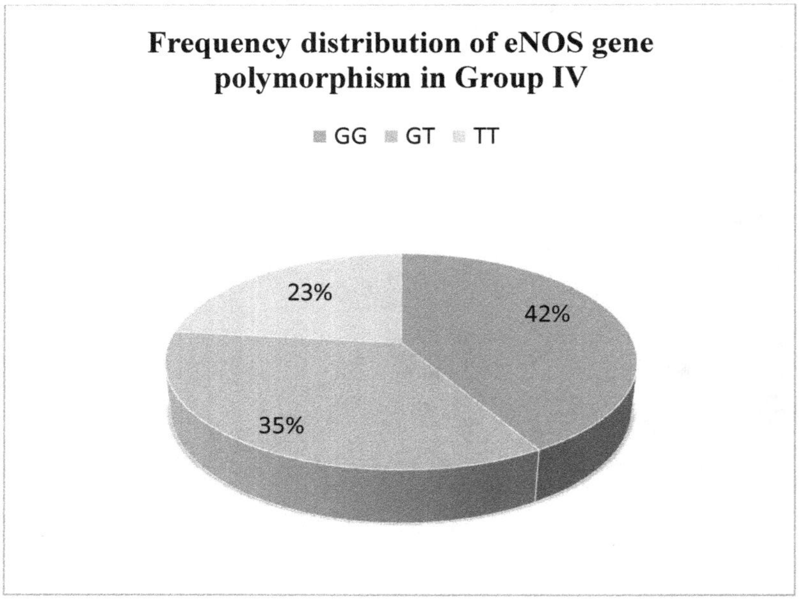

Chapter 7

Discussion

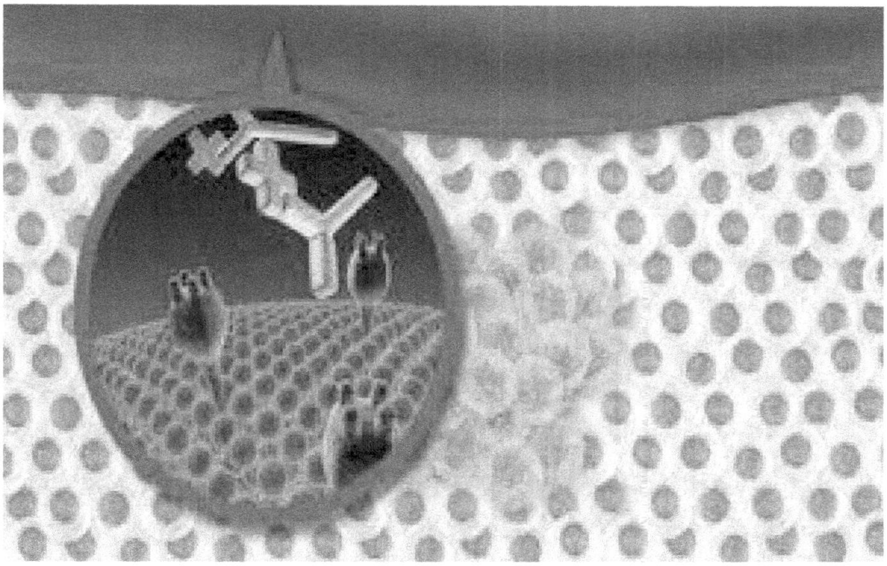

'Metabolic Syndrome' is a widely prevalent disorder that presents in a distinct, albeit heterogenous phenotype. Although obesity and insulin resistance are not synonymous with the metabolic syndrome, they are integral features in this derangement of adipocyte physiology and carbohydrate metabolism. It is now proposed that this syndrome predicts the development of type 2 diabetes mellitus and cardiovascular disease.[338]

Type 2 diabetes is a multi factorial disease with impaired insulin synthesis, release and action at various levels and through a number of mechanisms. It affects glucose homeostasis and reflects changes in many components in the network that regulates glucose homeostasis. Long term hyperglycemia accompanied by dyslipidemia and insulin resistance affects the vascular endothelium silently leading to early damage. Prolonged damage to endothelium leads to complications manifesting through eye, kidney, heart, nerve, or blood vessels.[339]

Therefore, the aim of this study was to assess the extent of endothelial dysfunction by analyzing endothelial nitric oxide synthase (eNOS) gene polymorphism in metabolic syndrome and type 2 diabetes mellitus with and without complications & correlate the same with phenotypic clinical parameters.

The study population consisted of 230 subjects which were divided in to four groups based on the inclusion criteria. The four groups were composed of Group I (n=50) - Controls; Group II (n=60) - Metabolic Syndrome patients; Group III (n=60) - Type 2 Diabetic subjects; Group IV (n=60) - Type 2 Diabetic subjects with complications.

In all four groups, the following were studied:

- Athropometric measurements
- Glycemic indices
- Lipid parameters
- Endothelial markers
- eNOS gene polymorphism

Correlation of different parameters with standard model of insulin resistance such as HOMA IR was done. The association of single nucleotide polymorphism in eNOS gene with risk factors of metabolic syndrome and type 2 diabetes was analyzed.

Insulin Resistance

Insulin resistance is a pathophysiological condition in which normal insulin concentration does not adequately produce a normal insulin response in target tissues such as adipose, muscle and liver. Under this condition, pancreatic beta cells secrete more insulin (i.e. hyperinsulinemia) to overcome the hyperglycemia in these individuals. Over time, the inability of the pancreatic beta cells to produce sufficient insulin to correct worsening tissue insulin resistance leads to hyperglycemia and overt type 2 Diabetes.[340]

Evidence for the central role of insulin resistance in development of metabolic syndrome is supported by the Bruneck study which examined the prevalence of insulin resistance among 4800 subjects aged 40–79 years, using the homeostasis model assessment (HOMA) method. In this study, the degree of insulin resistance correlated with the number of metabolic abnormalities and when several abnormalities were clustered together, insulin resistance was almost always present. The link between insulin resistance and CVD is probably through endothelial cell dysfunction, causing vascular damage.[341]

Literature review shows that a variety of methods of varying complexity are used to estimate insulin resistance and each of these methods has its own advantages and disadvantages. [198, 342] The choice of method depends on the size and type of study being conducted and no single method is appropriate under all circumstances. The hyperinsulinemic-euglycemic clamp method is the gold standard for measuring insulin sensitivity [342] but it is impractical for routine use, expensive, time-consuming and labor-intensive[343] compared to homeostatic measurements.[342, 344] Currently, the homeostasis model of assessment (HOMA) which correlates well with gold standard euglycemic clamp techniques, is widely used in clinical and epidemiological research because of its simplicity. The original HOMA1 model was first described in 1985 by Matthews et al [185] and has been used in many longitudinal and epidemiological studies to derive β-cell function (%B) and insulin resistance values from fasting plasma insulin (FPI) and fasting plasma glucose (FPG) with the formula mentioned earlier. Though HOMA IR is the widely used validation measure for insulin resistance, it is expensive and is mainly used for clinical research purposes. For daily clinical practice, it is necessary to use other easily applied measurements in the general population.

Elevated WHR, BMI are the factors which determine body fat distribution and may contribute to insulin resistance state hence they were correlated with HOMA IR in our study. As association exists between dyslipidemia and insulin resistance, one of the indices that has been used to evaluate insulin resistance is the ratio of TG/HDL-C. It has been observed that individuals with metabolic syndrome show altered lipid metabolism such as elevated concentrations of TG-rich lipoproteins and other cholesterol rich particles.[345] In addition; triglyceride content in liver and muscle, [346-352] is a determining factor for development of insulin resistance. These findings support the important role played by triglycerides in insulin resistance. *Hence TG/HDL-C ratio was calculated in the control and study groups as an indicator of insulin resistance*.

Insulin resistance is also characterized by overproduction of Apo B containing lipoproteins, metabolic dyslipidemia, (hypertriglyceridemia, high plasma VLDL, low amount of HDL cholesterol),[353] small dense LDL [354] and ultimately the development of cardiovascular disease.[355] Insulin also regulates Apo B degradation indirectly through mitogen activated protein. Insulin resistance on the other hand, decreases Apo B protein degradation resulting in increased levels of Apo B. *Hence Apo B levels were also estimated to study insulin resistance.*

7.1 Anthropometric parameters

MS is represented by a cluster of risk factors which can also be present in type 2 diabetes with and without complications. The factors of MS like dyslipidemia, central obesity are associated with an increased body mass index (BMI) and waist-to-hip ratio (WHR).[356]

The anthropometric parameters measured in this study were weight, height, waist and hip circumference and blood Pressure. *BMI and WHR were calculated for all the controls and subjects.* The descriptive comparison is shown in table 6.1.

The most direct measures of body fat such as underwater weighing or dual-energy x-ray absorptiometry (DEXA) scanning are impractical for use.The recent techniques like

Magnetic Resonance Imaging (MRI) and Spectroscopy (MRS) have emerged as the gold-standard for accomplishing this task but these techniques are expensive and not always practical.[357]

Indirect estimates of body fat are more clinically practical in routine use. These indirect estimates included body mass index (BMI) which is obtained by dividing the body weight in kilograms by height in meters squared, or fat distribution indices like waist-hip ratio and waist circumference (WC).[358]

BMI, WC, were taken as risk factors for metabolic syndrome and type 2 diabetes as per WHO and NCEP ATP III criteria.[9,78] These measures were used as a simple method to determine body fat distribution.[359] The information obtained from this study was useful for screening purposes and correlated with HOMA IR to understand the underlying mechanism of insulin resistance.

Table 6.1 represents anthropometric parameters in study groups and control group. In all the study groups BMI and WHR was significantly higher as compared to control but no significant difference was found when comparison was done among the study groups as shown in table 6.1a, 6.1b, 6.1c, 6.1d, 6.1e, 6.1f.

7.1.1 Comparison of BMI and WHR among different study groups and their correlation with HOMA IR

i) Group I (Control) and Group II (MS)

When BMI and WHR, the indicators of insulin resistance were compared among Group I and Group II, we found both were significantly higher in Group II ($p<0.0001$) which is presented in table 6.1a. HOMA IR was correlated with WHR by univariate analysis and positive correlation was obtained with r value=0.28 ($p<0.05$). The correlation graph is presented as figure 6.1a. Simple regression analysis was also performed between HOMA IR and BMI of Group II as can be seen from figure 6.1d. The correlation analysis showed positive correlation between these two variables, r=0.48, ($p<0.01$).

The combined multivariate analysis was performed taking HOMA IR as dependent variable and WHR and BMI as independent variables. WHR showed statistically significant relation whereas BMI correlated positively with HOMA IR but was less significant ($p<0.05$) as compared to WHR and can be seen from table 6.1g.

Our results of WHR and BMI suggest that WHR is a better indicator of insulin resistnce in metabolic syndrome patients which is contradicted by the study conducted by Gerald Reaven in 2005, according to whom BMI is a predictor of insulin resistance.[360]

Our findings of WHR are in accordance with Arnlov J. et al who concluded that middle-aged men with metabolic syndrome or insulin resistance had an increased risk of type 2 diabetes, regardless of BMI status during 20 years of follow-up compared with normal weight men without metabolic syndrome or insulin resistance.[361]

As per our findings that WHR and BMI remained significant predictors of metabolic syndrome is in accordance with that of Han et al.[362]

ii) Group I (Control) and Group III (T2DM)

There was statistically significant difference between WHR and BMI between Group I and Group III ($p<0.0001$) as shown in table 6.1b. Correlation analysis of Group III showed positive between HOMA IR and WHR ($r=0.57$, $P< 0.01$). HOMA IR and BMI also correlated positively ($r=0.78$, $p< 0.01$). The correlation is shown in figure 6.1b, 6.1e and table 6.1j, 6.1k respectively. The result of combined regression analysis between HOMA IR and two variables; BMI and WHR showed the statistical significance and is presented in table 6.1h.

A meta analysis done by Vazquez et al [363] showed that body mass index, waist circumference and WHR are associated with type 2 diabetes which is similar to our results for Group III.

The regression model performed by taking HOMA IR as dependent variable and WHR and BMI as independent variables showed that both the variables showed positive correlation which is statistically significant, as shown in table 6.1h.

A study conducted by Mathur et al in Asian North Indians to assess whether insulin resistance is an important causative factor underlying type 2 diabetes, proved that diabetics had significantly higher WHR.[364]

iii) Group I (Control) and Group IV(T2DM with complications)

Table 1c depicts comparison of WHR and BMI of Group I with Group IV. There was a positive correlation when simple regression analysis of HOMA IR with WHR was done ($r=0.44$, $p< 0.01$). There was also positive correlation between HOMA IR and BMI but not statistically significant ($r=0.24$) which is presented in figure 6.1c, 6.1f and table 6.1j, 6.1k. The combined regression model of HOMA IR with WHR and BMI in Group IV showed significant positive correlation only with WHR which is presented in table 6.1i.

iv) Group II (MS) with Group III (T2DM) and Group IV (T2DM with complications)

v) Group III (T2DM) and Group IV (T2DM with complications)

The comparison of WHR and BMI between Group II and III; Group II and IV as well as between Group III and IV were statistically insignificant and presented in table 6.1d, 6.1e, 6.1f.

Thus the finding of anthropometric parameters in all the study groups indicated that there was good level of correlation between BMI and WHR as indices of insulin resistance. Measurement of Body Mass Index has been widely considered as an index of central obesity and a marker of body fat due to convenience and cost effectiveness.

WHR is closely correlated to all of the components of the metabolic syndrome. This was first demonstrated in 1991 in a study that compared Europeans with South Asians and examined a group of 3193 men and 561 women aged between 40–69 years in London. In comparison with the European subjects, the south Asian subjects had a higher prevalence of diabetes (19 vs. 4%).[365] Das et al in their study showed statistically significant difference in WHR and BMI values in type 2 diabetes compared to control group. They also observed significant positive correlation between HOMA IR, BMI and WHR. Our findings of WHR and BMI in study groups are in accordance with Das et al.[366]

Multivariate regression analysis of WHR ratio and BMI with HOMA IR in all the study groups showed that WHR ratio correlated well with HOMA IR in all the groups in comparison with BMI and these findings are in accordance with Chen et al that waist to hip ratio is a better predictor of insulin resistance in metabolic syndrome and type 2 diabetes.[295]

Conclusion

Elevated WHR and BMI which are the factors of metabolic syndrome were measured and correlated with insulin resistance in all the study groups using univariate and multivariate regression analysis. Both the indices showed good degree of correlation.

7.2 Blood Pressure

High blood pressure is a classical feature of metabolic syndrome. It is also known that metabolic syndrome is present in one third of hypertensive patients.[367,368] Hence, high blood pressure is included as part of definition for metabolic syndrome by World Health Organization, National Cholesterol Education Program, International Diabetes Federation and the American Heart Association/National Heart, Lung and Blood Institutes.[9,356,369,370]

Both systolic and diastolic blood pressure (Systolic and diastolic) differed statistically in all study groups on comparing with control ($p<0.0001$). The comparison within the groups also showed statistically significant difference.

7.2.1 Comparison of Blood pressure among different study Groups

i) Group I (Control) and Group II (MS)

When systolic and diastolic blood pressure values were compared between Group I and Group II, we found significantly higher values in Group II ($p<0.0001$) as shown in table 6.1a

ii) Group I (Control) and Group III (T2DM)

The comparison of systolic and diastolic blood pressure between Group I and Group III showed statistically significant difference ($p<0.0001$), which is presented in table 6.1b.

iii) Group I (Control) and Group IV (T2DM with complication)

There was statistically significant difference between both systolic and diastolic blood pressure between Group I and Group IV ($p<0.0001$), refer table 6.1c.

iv) Comparison of Blood pressure of Group II with Group III and Group IV

The comparison of systolic and diastolic blood pressure between Group II and Group III was marginally high which was statistically insignificant ($p<0.09$), as shown in table 6.1d whereas the difference in blood pressure of Group II and IV was statistically significant ($p <0.0001$) as presented in table 6.1e.

v) Comparison of Blood pressure ratio between Group III and Group IV

Table 6.1f presents difference in systolic and diastolic blood pressure of Group III and IV ($p<0.0001$) which was significant.

Blood pressure levels are strongly associated with visceral obesity and insulin resistance[371] which are the main pathophysiologic features underlying the metabolic syndrome. Several mechanisms connect insulin resistance with hypertension in the metabolic syndrome. An anti-natriuretic effect of insulin has been established by accumulating data indicating that insulin stimulates renal sodium re-absorption.[178,372,373] This antinatriuretic effect is preserved and may be increased in individuals with insulin resistance. This effect may play an important role for development of hypertension in the metabolic syndrome.[374] In insulin resistant state the normal response of insulin-mediated glucose disposal and incremental blood flow in response to insulin is lost, suggesting resistance to the action of insulin to induce vascular NO production.[375] There is increasing evidence that angeotesin II inhibits insulin and IGF-1 signaling through the PI3K/Akt pathway resulting in inhibition of mechanisms involved in the vasodilator and glucose transport resulting in endothelial dysfunction.[376-378] These factors may induce sympathetic over activity, vasoconstriction, increased intravascular fluid and decreased vasodilatation, leading to development of hypertension in metabolic syndrome.

Our results are in agreement with the study by Lu et al and Chen et al who reported that hypertension is linked to metabolic syndrome and risk factors of metabolic syndrome.[290,379]

Conclusion

The systolic and diastolic blood pressure values were statistically significant in all the study groups supporting the fact that hypertension is associated with insulin resistance, one of the pathophysiologic features underlying metabolic syndrome.

7.3 Glycemic Parameters

Measurement of glycemic parameters included fasting & post prandial plasma sugar, HbA1c and fasting plasma insulin in all the groups which is presented in table 6.2 and also as graphs 6.2a and 6.2b. HOMA IR was calculated using the formula mentioned earlier in chapter 5.

7.3.1 Comparison of FPG, PPPG, FPI, HbA1c and HOMA IR among the study Groups

i) Group I (Control) and Group II (metabolic syndrome)

Healthy individual were enrolled in Control Group. Patients were screened for fasting plasma glucose and those with sugar value ≥110 mg/dl as per NCEP ATP III criteria were enrolled in Group II. The comparison of FPG between Group I (control) and Group II (MS) was significantly high as shown in table 6.2a ($p<0.0001$). The values of the fasting plasma glucose indicated impaired fasting glucose in all the MS patients.

Impaired fasting glucose (IFG) and impaired glucose tolerance (IGT) represent intermediate states of abnormal glucose regulation that exist between normal glucose homeostasis and diabetes. IFG is defined by an elevated fasting plasma glucose (FPG) concentration (≥110 and ≤126 mg/dl).[380] IGT is defined by an elevated 2-hrs plasma glucose concentration after a 75-g glucose load on the oral glucose tolerance test (OGTT) in the presence of an FPG concentration ≥126 mg/dl). [380,381] It should be typically ≥140 mg/dl and ≤ 200 mg/dl.

There is no clear distinction between IFG and IGT from a pathophysiologic perspective. IFG is suggested to be mainly characterized by insulin resistance in the liver leading to increased basal glucose output and by insufficient basal insulin production from the pancreas to control glucose levels in the fasting state. IGT is characterized by defective insulin action in peripheral tissues, mainly in the muscles, leading to defective glucose uptake and also in the liver with defective control of hepatic glucose output, as well as defective insulin secretion after carbohydrate absorption.[382] The natural history of both IFG and IGT can be defined in terms of progression to diabetes. As discussed earlier, the majority of people with IFG/IGT will develop progressive hyperglycemia and eventually diabetes.

HbA1c was estimated in all the groups. The values of HbA1c in Group I and Group II were within the normal limits but difference in these two groups is statistically significant.

Our results of HbA1c are in accordance with Park et al who demonstrated its increased percentage and established cut off for HbA1c. He also suggested that HbA1c might be used as a diagnostic criterion for metabolic syndrome.[383]

Fasting plasma insulin (FPI) was estimated in all groups for measurement of insulin resistance by Homeostatis Model of Assessment of insulin resistance (HOMA IR). Fasting insulin levels were raised in metabolic syndrome Group in comparison with control Group ($p<0.01$) as per table 6.2. The HOMA IR in MS Group was elevated showing the state of insulin resistance, which may be due to increased BMI, WHR, FPG and dyslipidemia.

The findings of HOMA IR and FPI in MS Group showed that Group II subjects presented insulin resistance state and may be at a high risk for the development of diabetes. Risk is particularly high when glucose dysregulation is present (IFG or IGT). Framingham Heart data found that the relative risk for diabetes was five-fold higher in patients with metabolic syndrome.[370]

Our results of FPI are similar to the study carried out by Lorenzo C et al who found elevated levels of FPI in patients with metabolic syndrome.[43]Haffner et al showed that pre diabetic subjects have an atherogenic pattern of cardiovascular risk factors and these changes are predominantly observed in pre diabetic subjects with increased HOMA IR and fasting insulin ("insulin resistance") at baseline.[282]

ii) Group I (Control) & Group III (T2DM)

Group III comprised of type 2 diabetes patients with fasting blood glucose level ≥ 126 mg/dl as per WHO/ICMR criteria. As shown in table 6.2b the fasting and postprandial blood glucose levels were significantly higher in Group III on comparing with control Group ($p<0.0001$). The HbA1c values also differed significantly in Group III ($p<0.0001$). The fasting plasma insulin levels and HOMA IR were elevated in comparison to control ($p < 0.0001$).

Our findings of glycemic parameters in Group III are similar with the findings of Simoes et al who reported high BMI, Hyperglycemia and elevated HbA1c in type 2 diabetic patients.[384]

Bodhe et al also demonstrated high fasting blood glucose level and HbA1c percentage in type 2 diabetic patients.[385]

Our results for insulin levels & HOMA IR values for Group III are also in accordance with Okita et al who reported that HOMA-IR correlated with various clinical parameters in patients with poorly controlled type 2 diabetes after administration of insulin. Their results suggested that HOMA-IR is a reliable and useful parameter for evaluation of insulin sensitivity in patients with type 2 diabetes treated with insulin.[310]

Our finding that HOMA IR is an indicator of insulin resistance in type 2 diabetes with high BMI and WHR is in accordance with Yokoyama et al,[285] who quantified insulin sensitivity check index (QUICKI), HOMA IR and the reciprocal index of homeostasis model assessment (1/HOMA-IR) in moderately obese type 2 diabetic and healthy subjects. They reported that these are excellent surrogate indices for insulin resistance.

iii) Group I (Control) & Group IV (T2DM with complications)

Group IV comprised of type 2 diabetic patients with complications. As shown in Table 6.2c, the fasting and postprandial blood glucose levels were significantly high in Group IV when compared with control Group ($p<0.0001$). The HbA1c values were also significantly raised in Group IV ($p<0.0001$). The fasting insulin levels and HOMA IR were elevated in comparison to control and were statistically significant ($p < 0.0001$).

Murugan et al[386] in their study on diabetics, observed hyperglycemia and increased percentage of HbA1c similar to our findings and suggested that though blood glucose is considered as a prime test for optimizing treatment of diabetes mellitus, determination of HbA1c is a better to long term glucose control for patient management and is irrespective of daily glucose levels. This would prevent or delay further the diabetic complications.

Disturbed glucose metabolism, which is a common cause of hyperglycemia, increased insulin secretion and resistance to insulin was observed in all the study groups.

iv) Group II (MS) with Study Groups III (T2DM) and IV (T2DM with complications)

v) Group III (T2DM) with Group IV (T2DM with complications)

Glycemic parameters of Group II were compared with Group III and IV as shown in table 6.2d and 6.2e. The FPG, PPPG and HbA1c were significantly higher in both the Groups when compared with Group II. The values of FPI were significantly lower and HOMA IR was significantly higher in Group III and IV in comparison with Group II as shown in table 6.2f.

There is striking difference in the values of FPG, PPPG, HbA1c, FPI and HOMA IR of Group III and Group IV than those of Group II.

The comparison of glycemic parameters between Group III and Group IV also showed that FPG and PPPG were significantly higher whereas FPI was lower significantly. HbA1c and HOMA IR did not show any significant difference.

This typical picture of glycemic parameters in Groups II, III and IV shows that development of type 2 diabetes is characterized by presence of two fundamental defects, i.e. insulin resistance and impaired insulin secretion which disrupt the delicate balance between the two by which target tissues for insulin communicate with the beta cells and vice versa. Thus, type 2 diabetes is a heterogeneous disorder characterized by impaired insulin secretion, diminished peripheral (muscular) insulin action and increased hepatic glucose production along with beta cell dysfunction.[387]

Impaired insulin secretion is found uniformly in type 2 diabetic patients in all ethnic populations.[49, 388, 389-398] Early in the development of type 2 diabetes, insulin resistance was well established but glucose tolerance remains normal because of a compensatory increase in insulin secretion. This dynamic interaction between insulin secretion and insulin resistance has been well documented. Within the normal glucose tolerant population, approximately 20%–25% of individuals are severely resistant to the stimulatory effect of insulin on glucose uptake and subjects in the lowest quartile of insulin sensitivity are as insulin resistant as type 2 diabetics. Insulin secretion in these insulin-resistant, non diabetic individuals, however, is increased in proportion to the severity of the insulin resistance and glucose tolerance remains normal. Thus, the beta cells are able to ''read'' the severity of insulin resistance and adjust their secretion of insulin to maintain normal glucose tolerance. In type 2 diabetics, the fasting plasma insulin concentration is normal or increased and basal insulin secretion is elevated. The relationship between the fasting plasma glucose (FPG) and insulin concentrations resembles an inverted U shape or horseshoe.[388, 49]

As the fasting glucose rises from 80 to 140 mg/dl, the fasting plasma insulin concentration increases progressively, reaching a peak value 2.0–2.5-fold greater than in normal weight, non diabetic, age-matched controls. The progressive rise in fasting plasma insulin levels can be viewed as an adaptive response of the pancreas to offset the progressive deterioration in glucose homeostasis. When the FPG exceeds 140 mg/dl, the beta cells are unable to maintain its elevated rate of insulin secretion and the fasting insulin concentration declines precipitously.

A typical type 2 diabetic subject with a FPG level of 150–160 mg/dl secretes an amount of insulin similar to that in a healthy non diabetic individual; however, a ''normal'' insulin response in the presence of hyperglycemia and underlying insulin resistance is markedly abnormal. At FPG levels in excess of 150–160 mg/dl, the plasma insulin response, when viewed in absolute terms, becomes insulinopenic.[399]

Finally, when the fasting glucose exceeds 200–220 mg/dl, the plasma insulin response to a glucose challenge is markedly blunted. Nonetheless, the fasting hyperinsulinemia persists despite FPG concentrations as high as 250–300 mg/dl, and 24-hour integrated plasma insulin and C-peptide profiles in lean type 2 diabetic patients remain normal. These normal day-long values result from the combination of elevated fasting and decreased postprandial insulin and C-peptide secretory rates.[117] It should be emphasized that, even though the plasma insulin response is increased in absolute terms early in the development of type 2 diabetes (FPG $\leq$ 140 mg), this does not mean that beta-cell function is normal.

Once hyperglycemia exists, beta-cell dysfunction is clearly present in subjects with type 2 diabetes. This change manifests in a number of different ways including decline in the ability of glucose to potentiate the insulin response to non-glucose secretagogues.[400]

Recent examination of a number of different parameters in various human models has emphasized the fact that beta-cell dysfunction, when assessed using appropriate methods, is present well before the onset of hyperglycemia, i.e. in subjects with impaired glucose tolerance and even in first-degree relatives with normal glucose tolerance. First-phase insulin response to glucose is reduced in an early stage, thus contributing to postprandial hyperglycemia, while late-phase insulin secretion may be preserved for a longer term as it is triggered by sustained hyperglycemia.

A backward extrapolation of findings made in the United Kingdom Prospective Diabetes Study (UKPDS) cohort from shortly after the clinical diagnosis of type 2 diabetes, strongly suggests that beta-cell dysfunction commences years (at least 10 years) before hyperglycemia develops.[401]

Thus, subjects with type 2 diabetes are characterized by both tissue insulin resistance and impaired insulin secretion.[402, 403] The development of diabetes requires the presence of these two fundamental defects, which disrupt the delicate balance by which insulin-target tissues communicate with the beta cells and vice versa. To observe the development of frank diabetes mellitus, a defect in insulin secretion must be superimposed on insulin resistance[402, 404] and numerous observations suggest that this beta-cell defect might be genetically determined.[405]

Conclusion

Our finding of glycemic parameters indicated presence of insulin resistance in all the study groups which was measured in the form of HOMA IR.

7.4 Lipid parameters

In lipid parameters we estimated serum total cholesterol (TC), Triglycerides (TG) and HDL cholesterol in control and all study groups. LDL cholesterol and VLDL cholesterol values were calculated using Friedwald's formula. Apo B levels were measured using immuno turbidometric method. The data is presented in table 6.3 and graphs 6.3a, 6.3b resp.

7.4.1. Comparison of Lipid Parameters among study groups and correlation with HOMA IR

i) Group I (Control) and Groups II (MS)

Total cholesterol, TG, LDL cholesterol, VLDL and Apo B levels were significantly higher in Group II as compared to Group I and are shown in Table 6.3a ($p < 0.0001$). HDL cholesterol levels were significantly decreased in Group II on comparing with Group I ($p<0.0001$).

In our study we used HOMA IR, the widely accepted model for assessment of insulin resistance and compared with both TG/HDL-C and Apo B, an atherogenic marker of insulin resistance. Since TG/HDL-C and Apo B are indicators of insulin resistance, we assessed their relationship with HOMA IR.

The TG/HDL-C ratio was calculated in metabolic syndrome patients and Control Group and we found that TG/HDL-C ratio was higher in the study Groups. There was significant positive correlation between HOMA IR and TG/HDL-C ratio ($r=0.29$; $p<0.02$) in the study group which is presented in figure 6.3a. The regression model showed R^2 value of 0.087, indicating 8.7 % variation in the HOMA IR due to TG/HDL-C ratio as shown in table 6.3j.

The predictive level and significance of correlation between HOMA IR and Apo B was assessed by regression analysis considering HOMA IR as dependent variable. A significant positive correlation ($r=0.41$; $p<0.0009$) was obtained between HOMA IR and Apo B in metabolic syndrome patients as shown in figure 6.3d. The R^2 value showed by regression model was 0.17 showing 17 percent variation in HOMA IR due to Apo B levels, shown in table 6.3k.

When combined multivariate analysis of TG/HDL-C and Apo B with HOMA IR was performed, the regression model indicated that TG/HDL-C ratio is a non significant ($p>0.05$) variable whereas Apo B was statistically significant ($p<0.01$) as shown in table 6.3g.

Elevated triglycerides and low levels of HDL cholesterol characterize dyslipidemia in metabolic Syndrome. Hypertriglyceridemia is associated with alterations, both in structure and metabolism of HDL and LDL.[406] The presence of low HDL cholesterol in metabolic Syndrome is partially due to high triglycerides. As triglycerides increase, the cholesterol esters of HDL are exchanged for triglycerides through the action of cholesteryl ester protein, causing the HDL particle to become smaller and less dense. Through mechanisms not well understood, the smaller HDL particles is metabolized and cleared at an abnormally high rate, resulting in low HDL levels. This makes the smaller particles less antiatherogenic. The larger and more buoyant HDL particles can remove more free cholesterol they can remove from cells and atherosclerotic plaque.[406]

A common finding about LDL cholesterol levels is that LDL particles are smaller and denser than normal; a feature that is associated with increased cardiovascular risk.[407] Smaller LDLs also increases the ability to penetrate the vascular endothelium and a greater affinity for intimal glycans. The retention in the intima allows for enhanced oxidative transformation, endovascular injury and the accumulation of foam cells.[406] Most researchers agree that the larger the number of smaller, denser LDL particles, greater the cardiovascular risk.[370] In addition to increased numbers of smaller and denser LDL cholesterol particles, there is a corresponding increase in apolipoprotein B. In fact, it appears that once apolipoprotein B concentrations are elevated, cardiovascular risk can be considered high.[408]

Apo B is a potentially atherogenic lipoprotein, including lipoprotein-a (Lpa), LDL, intermediate density lipoproteins (IDL), VLDL and Triglyceride rich lipoprotein (TRL) remnants. Each of these lipoprotein particles contains one Apo B molecule. Apo B provides a direct measure of the number of atherogenic particles present in the circulation.[409, 410] A direct link between Apo B and severity of CVD in patients undergoing diagnostic cardiac catheterization was followed by numerous studies that supported Apo B as being highly predictive of CVD and in some cases, more closely linked to CVD outcomes than LDL-C.[409, 411,412]

Our values of TG/HDL-C are in accordance with the study of Salazar et al, [307] Our values for complete lipid profile are in line to the studies carried out by Maheria et al and Lorenzo et al [413,414]

Our results are also in accordance with the study of Morato et al [415] and Brehm et al [416] who demonstrated positive correlation between TG/HDL-C ratio and insulin resistance, confirming that TG/HDL-C ratio predicts insulin resistance in metabolic syndrome. But an independent study by Knights et al [417] stated that, triglyceride/high-density lipoprotein cholesterol ratio fails to predict insulin resistance in African American women.

Our results of Apo B in metabolic syndrome patients are matching with the results obtained by Sung et al and Pie et al. [418,419]

ii) Group I (Control) and Groups III (T2DM)

Table 6.3b shows that all lipid profile parameters including Apo B except for HDL were significantly increased in Group III as compared to Group I ($p < 0.0001$). Simple regression analysis was performed considering HOMA IR as a dependent variable and TG/HDL-C and Apo B as independent variables separately. Positive correlation was obtained between HOMA IR and TG/HDL-C ratio as well as HOMA IR and Apo B , ($r= 0.52$, $P<0.01$ and $r= 0.59$, $P<0.01$) respectively and shown in table 6.3j and 6.3k. The graphical presentation is done in figure 6.3b and 6.3e resp. The R^2 value of HOMA IR and TG/HDL-C was 0.28 indicating 28 percent variation in HOMA IR due to high TG/HDL-C ratio whereas HOMA IR with Apo B showed 35 percent ($R^2=0.35$) variation in HOMA IR due to elevated Apo B levels. The combined regression analysis proved that Apo B is better indicator of HOMA IR than TG/HDL-C ratio depicted in table 6.3h.

The main cause for the three cardinal features of diabetic dyslipidemia is increased free fatty-acid release from insulin-resistant fat cells.[420,421] The increased flux of free fatty acids into liver in the presence of adequate glycogen stores promotes triglyceride production, which in turn stimulates the secretion of apolipoprotein B (ApoB) and VLDL cholesterol. The impaired ability of insulin to inhibit free fatty-acid release leads to enhanced hepatic VLDL cholesterol production, [422] which correlates with the degree of hepatic fat accumulation.[423]

Hyperinsulinemia is also associated with low HDL cholesterol levels.[424] The increased number of VLDL cholesterol particles and increased plasma triglyceride levels decreases the level of HDL cholesterol and increases the concentration of small dense LDL-cholesterol particles via several processes: VLDL-transported triglyceride is exchanged for HDL-transported cholesteryl ester through the action of the cholesteryl ester transfer protein (CETP), which results in increased amounts of both atherogenic cholesterol-rich VLDL remnant particles and triglyceride-rich, cholesterol-depleted HDL particles.

Our results for Total Cholesterol, TG, HDL, VLDL, LDL and Apo B in type 2 diabetic group were decreased as shown by Sert et al [425] & Kanani et al.[426]

Our finding of increased Apo B in type 2 diabetes patients is in accordance with that of Jiang et al who prospectively followed 746 diabetic men in the Health Professionals follow-up study who were aged between 46–81 years and free of CVD at the time of blood collection. During 6 years follow up, they ascertained 103 incident CVD cases and concluded that non-HDL cholesterol and Apo B are more potent predictors of CVD incidence among diabetic men than LDL cholesterol.[427]

In Group III, statistically significant increase in Apo B levels were seen as well as its positive correlation with HOMA IR. Mehta et al in their study, evaluated usefulness of insulin resistance estimation and the metabolic Syndrome in predicting coronary atherosclerosis in type 2 diabetes mellitus and demonstrated that apolipoprotein B (apoB) and non–high-density lipoprotein cholesterol are associated with coronary artery calcification in DM. Multivariate analysis performed by him using traditional risk factors, Apo B and homeostasis model assessment of insulin resistance (HOMA-IR) were independently associated with coronary artery calcification in DM. In nested models, HOMA-IR correlated well with Apo B and vice a versa suggesting that estimates of insulin resistance adds value to MS and Apo B in predicting coronary artery calcification scores in DM.[303]

iii) Group I (Control) and Group IV(T2DM with complications)

Table 6.3c showed that all lipid profile parameters except for HDL and Apo B were significantly increased in Group IV as compared to Control Group ($p < 0.0001$). A positive correlation was obtained on performing simple regression analysis between HOMA IR, a dependent variable and TG/ HDL-C ratio as well as Apo B as independent variables in different models. The correlatin is shown in figure 6.3c and 6.3f. The correlation between HOMA IR and TG/HDL-C ratio was statistically significant(r=0.59 ;P<0.01) with R^2 value of 0.37 showing 37 percent variation in HOMA IR due to TG/HDL-C ratio. Another model between HOMA IR and Apo B also showed statistically significant correlation (r=0.61; P<0.01) with R^2 0.37 value indicating 37 percent variation. Refer table 6.3j and 6.3k for r and R^2 values. The combined multivariate analysis shown in table 6.3i revealed statistically significant correlation of HOMA IR with TG/HDL-C as well Apo B, Apo B correlated strongly with HOMA IR.

Dyslipidemia is one of the contributing factors for diabetic complications which includes micro and macrovascular complication, retinopathy and neuropathy as can be seen by our finding of positive correlation between HOMA IR and TG/HDL-C ratio. This observation is also endorsed by the study of Lee et al who demonstrated that high TG/HDL-C ratio was an important risk factor for nephropathy in type 2 diabetic subjects.[428]

Zoppini et al suggested in their study that the TG/HDL-C ratio was associated with an increased incidence of microvascular complications in individuals with type 2 diabetes mellitus without prior cardiovascular disease.[429]

Our finding about Apo B are in accordance with that of Sasongko et al who observed that Apo B ratio was associated with an increased incidence of microvascular complications in individuals with type 2 diabetes mellitus.[430,431] Patel et al stated in their study that atherogenic lipoproteins may play significant role in development of diabetic nephropathy.

Diabetic patients with most severe hyperglycemia have the highest risk of microangiopathy. However, the inverse relation may not be true. Hypertension, hypercholesterolaemia, dyslipidemia are additional major causes of microangiopathy. Risk of macroangiopathy does not appear to be strongly related to hyperglycemia, but is related to general risk factors for atherothrombosis. The cardiovascular risk factors such as hypertension, dyslipidemia, insulin resistance, hyperinsulinemia and impaired fibrinolysis cluster in the metabolic syndrome. All of the above mentioned factors create a state of constant and progressive damage to the vascular wall and cause endothelial dysfunction.[432] Dysfunction of the vascular endothelium is regarded as an important factor in the pathogenesis of micro- and macro-angiopathy.[433, 434]

iv) Group II (Control) with Group III (T2DM) and Group IV(T2DM with complications)

Comparison of all lipid parameters viz. total cholesterol, triglycerides, VLDL, LDL and Apo B showed significant increased ($P<0.0001$) where as HDL cholesterol was significantly decreased ($p<0.05$) in Group III and Group IV subjects on comparing with Group II subjects. Refer to table 6.3d & 6.3e.

v) Group III (Control) with Group IV (T2DM with complications)

All Lipid parameters showed insignificant difference when comparison was done between Group III and IV whereas Apo B was significantly increased in Group IV in comparison to Group III, shown in 6.3f ($p<0.01$).

Conclusion

- *TG/HDL-C and Apo B which are correlated to insulin resistance correlated well with HOMA IR in all study groups.*
- *Increased total cholesterol, low HDL and elevated TG/HDL-C ratio were correlated with insulin resistance and thus hyperinsulinemia.*
- *Apo B, the atherogenic lipoprotein is elevated and correlated with insulin resistance in all study groups.*

7.5 Endothelial markers

Endothelial markers involved estimation of serum nitric oxide and urinary microalbumin in all the study group as shown in table 6.4. Serum nitric oxide levels were significantly decreased in all the study groups compared to control. The urinary microalbumin levels were increased in all the study groups on comparison with control. The steady state of renal endothelial functions and hemodynamic harmonization may be changed because of cascading reactions caused by insulin resistance.[299] To study this association between the HOMA-IR and serum nitric oxide, regression analysis was performed between HOMA IR and serum nitric oxide in one model. In another model HOMA IR was compared with urinary microalbuminuria levels in all the study groups.

7.5.1 Comparison of Endothelial markers among all the study Groups

i) Group I (Control) and Group II (MS)

Table 6.4a represents the values for serum nitric oxide and urinary microlabumin in Group I (Controls) and Group II subjects (MS). Serum nitric oxide levels were significantly decreased ($p<0.05$) as compared to control and urinary microalbumin levels were increased in metabolic syndrome patients as compared to control but not statistically significant. The regression model performed between HOMA IR and serum nitric oxide which showed statistically significant negative correlation ($r=-0.28$; $P< 0.05$) and R^2 value was 0.08 showing eight percent variation in HOMA IR due to serum nitric oxide levels. Another model of HOMA IR and urinary microalbuminuria showed statistically significant positive correlation ($r=0.41$;$P< 0.05$) with R^2 value as 0.17 i.e seventeen percent variation in HOMA IR due to urine microalbumiuria. Refer table 6.4j, 6.4k and figure 6.4a and 6.4d.

The combined regression analysis model between HOMA IR as a dependent variable and nitric oxide and urinary microlbumin as independent variables is shown in table 6.4g. The model shows statistically significant association.

Our results for microalbuminuria in Group II are in accordance with Palaniappan et al who investigated the association of microalbuminuria with metabolic syndrome both as a whole and its individual components. In their study they reported that microalbuminuria was strongly associated with hypertension in both men and women.[286]

Our finding of microalbuminuria observed in metabolic syndrome patients is in agreement with Sheng et al who investigated the prevalence of microalbuminuria and its association with metabolic syndrome and its components in Chinese population. They found that prevalence of microalbuminuria was 3.2 times higher in subjects with metabolic syndrome than without.[435] Krishnaveni et al also observed that microalbuminuria is significantly associated with other features of Metabolic Syndrome.[436]

ii) Group I (Control) and Group III (T2DM)

The comparison of endothelial markers in Group I and III is shown in Table 6.4b. The levels of Serum nitric oxide were significantly diminished in Group III ($p < 0.001$) and the values of urinary microalbumin were significantly increased ($p < 0.0001$). The regression analysis between HOMA IR and serum nitric oxide showed significant negative correlation ($r=-0.41; P< 0.01$) and R^2 value was 0.17, showing seventeen percent variation in HOMA IR which is presented in table 6.4j and figure 6.4b. Simple regression analysis between dependent variable HOMA IR and urinary microalbumin showed positive correlation which was statistically significant as shown in table 6.4k and figure 6.4e. ($r=0.49; P< 0.01$) and R^2 value was 0.24 influencing twenty four percent variation in HOMA IR, shown in table 6.4k and figure 6.4b. The multivariate regression analysis between HOMA IR as dependent variable, with microalbumin and nitric oxide as independent variables has shown statistically significant correlation ($p<0.01$), table 6.4h.

In our study we found significant positive correlation between HOMA IR and microalbuminuria which is endorsed by the study of Korah et al who evaluated the relationship between microalbuminuria and insulin resistance in patients with T2DM having poor glycemic control. They reported that microalbuminuria is associated with IR and related CV risk factors like diabetes duration, poor glycemic control, and hypertension.[437]

Hsu et al also assessed the association between insulin resistance and urinary microalbumin in type 2 diabetes and concluded that insulin resistance could significantly predict development of microalbuminuria in type 2 diabetic patients.[299]

Varghese et al determined the prevalence and risk factors for microalbuminuria among south Indian type 2 diabetic patients and found that prevalence of microalbuminuria increased with the increase in duration of diabetes.[438]

iii) Group I (Control) and Group IV (T2DM with complications)

Table 6.4d shows the result of urinay microalbumin and serum nitric oxide. Serum nitric oxide is significantly lowered and urinary microalbumin is significantly increased in Group IV subjects in comparison with Group I. ($p< 0.0001$).

The regression analysis between HOMA IR and serum nitric oxide showed significant negative correlation ($r=-0.47;P< 0.0001$) and R^2 value was 0.22, showing twenty two percent variation in HOMA IR. Simple regression analysis between dependent variable HOMA IR and urinary microalbumin showed positive correlation which was statistically significant ($r=0.56$ $P< 0.0001$) and R^2 value was influencing thirty one percent variation in HOMA IR. Refer table 6.4j, 6.4k and figure 6.4c, 6.4f resp.

When the multivariate regression analysis between HOMA IR as dependent variable, with microalbumin as well as nitric oxide as independent variables was performed, both the variables have shown statistically significant correlation ($P<0.01$) table 6.4i.

Manaviat et al in his study on the risk factors for development of retinopathy and microalbuminuria and their correlation in type II diabetic patients, reported that microalbuminuria is associated with diabetic retinopathy in type II diabetic patients and is a reliable marker of retinopathy.[439]

Rani et al[300] conducted a study of estimation of the prevalence of albuminuria (micro - and macroalbuminuria) among persons with type 2 diabetes and determination of risk factor for presence and severity of diabetic retinopathy. They found that every 6th individual in the population of type 2 diabetes is likely to have albuminuria. Subjects with microalbuminuria were around 2 times as likely to have diabetic retinopathy as those without microalbuminuria. This is in accordance with our finding that microalbuminuira is associated with diabetic complications.

iv) Group II (Control) with Group III (T2DM) and Groups IV (T2DM with complications)

Serum nitric oxide and urinary microalbumin levels significantly differed in Group III & IV subjects in comparison with Group II as shown in table 6.4d & 6.4e ($p<0.0001$).

v) Group III (Control) & Group IV (T2DM with complications)

Both endothelial parameters showed significant difference when comparison was done between Group III & IV ($p<0.05$), shown in table 6.4f. Endothelial dysfunction indicated by impaired endothelium-dependent vasodilation and abnormal plasma concentrations of markers of endothelium. It is common in early and otherwise uncomplicated type II diabetes.

Endothelial dysfunction merely represents the impact of hyperglycaemia and other features of the metabolic syndrome. Insulin resistance usually precedes the development of type II diabetes and is often accompanied by a cluster of other risk factors such as hypertension, high triacylglycerol levels, low HDL levels, abnormal LDL composition, hyperinsulinaemia and insulin resistance. The mechanisms underlying this clustering are still unclear, but all the elements of cluster share two important pathophysiological features, namely insulin resistance and endothelial dysfunction. Insulin is a vasoactive hormone. It increases muscle blood flow in a time and concentration-dependent manner through a mechanism that involves binding to the insulin receptor on the endothelial cell membrane. Nevertheless, insulin induced increase in glucose uptake and total blood flow have different concentration–effect curves as well as kinetics.

Secondly, in a process termed capillary recruitment, insulin can redirect blood flow in skeletal muscle from non-nutritive capillaries (those that are not coupled to muscle cells) to nutritive capillaries (those that are) and thus increase glucose disposal even without increasing total blood flow.[440-442] In this way, physical integrity and normal function of the arteriolar and capillary endothelium are prerequisites for normal metabolic insulin action. Indeed, insulin's vasodilator actions have been shown to be impaired in classic insulin-resistant states, notably type II diabetes and hypertension. It is also known to be decreased by mediators closely associated with insulin resistance, nonesterified fatty acids (NEFAs) and an impaired NO-dependent action.[443-446] The endothelial dysfunction in large arteries is an early and prominent event in atherothrombotic disease which is paralleled by endothelial dysfunction in resistant vessels and metabolically important capillary beds that contributes to the development of metabolic syndrome.

Endothelial dysfunction in diabetes originates from three main sources.[447-450] First, hyperglycaemia and its immediate biochemical sequelae directly alter endothelial function.[447] Glucose transport into endothelial and vascular smooth muscle cells occurs by facilitated diffusion and is thus insulin-independent. Glucose transport is autoregulated by glucose in smooth muscle cells, but not in endothelial cells, in which an increase in blood glucose concentration will thus increase the intracellular accumulation of glucose and its metabolites. Thus, endothelial cells exposed to high glucose *in vitro* increase the production of extracellular matrix components, such as collagen and fibronectin and of procoagulant proteins, such as vWF and tissue factor and show decreased proliferation, migration and fibrinolytic potential and increased apoptosis.[451-456] Secondly, high glucose influences endothelial cell functioning indirectly by the synthesis of growth factor and vasoactive agents in other cells.[457] Thirdly, the components of the metabolic syndrome can affect endothelial function.

In type 2 diabetes an increase in intracellular glucose will lead to an increase in the flux of glucose to sorbitol via the polyol pathway, an increase in glucosamine-6- phosphate via the hexosamine pathway and the activation of PKC (protein kinase C) via de novo synthesis of DAG (diacylglycerol). In addition, glucose and glucose derived dicarbonyl compounds react non-enzymatically with the basic amino acids lysine and arginine in proteins to form AGEs (advanced glycosylation end-products) both extra and intra-cellularly. These different pathways are interrelated and potentiate each other. This pathway may impair endothelial function through three mechanisms. First, increased sorbitol accumulation will increase osmotic stress. Sorbitol accumulation decreases other osmolytes such as *myo*-inositol and taurine. However, the relatively low expression of aldose reductase in endothelial cells may not be sufficient to cause significant sorbitol accumulation. Secondly, the increase in the cytosolic $NADH/NAD^+$ ratio results in a redox imbalance that resembles a state which occurs in tissue hypoxia and is therefore termed hyperglycaemic pseudohypoxia. [458] Thirdly, the redox imbalance favours the accumulation of triose phosphates which increases the formation of methylglyoxal and AGEs and enhances oxidative stress, which can be exacerbated by NADPH-deficiency-induced depletion of reduced glutathione.

The cellular pathogenic consequences of hyperglycaemia induced activation of PKC are multiple and include dysregulation of vascular permeability directly or indirectly the latter through the induction of VEGF (vascular endothelial growth factor) in smooth muscle cells[456] dysregulation of blood flow by decreasing endothelial NOS activity and (or) increasing ET-1 synthesis.[460]

Although the underlying mechanisms of endothelial NO dysfunction in type 2 diabetes have been intensively studied and various mechanisms responsible for decreased endothelial NO bioactivity under the disease condition have been suggested, no single mechanism can fully explain the endothelial dysfunction. This may simply be due to the fact that multiple regulatory mechanisms are involved in endothelial NO bioactivity, particularly at the eNOS enzymatic level. Pathophysiologically, endothelial NO bioactivity is simply determined by the balance between synthesis and degradation of the molecule. Biochemically, the endothelial NO production is regulated at three different levels: 1) eNOS gene expression, 2) eNOS enzymatic activity and 3) degradation of NO. Dysfunction of any of these mechanisms can cause endothelial dysfunction.[461]

Serum nitric oxide levels were decreased statistically in Group II, III and IV. The decreased levels of serum NO in type 2 diabetes is in accordance with Gosh et al[462]

We found decreased levels of serum nitric oxide in metabolic syndrome patients in comparison with healthy controls. The study conducted by Dichi et al showed decreased level of serum nitric oxide in metabolic syndrome patients.[463]

Another important endothelial marker is microalbuminuria which is a major component of the metabolic syndrome. Microalbminuria is due to increase intraglomerular pressure causing transvascular leakage of albumin. Intraglomerular pressure rise is due to increased systolic and diastolic pressure. Intraglumerular hypertension is caused by hyperinsulinemia which increases vascular smooth muscle proliferation and increased sodium reabsorption and water retention through direct effect of insulin on DCT. Increased trans membrane permeability is also due to increase in pore size and decrease in number of negative charges in membrane because of thickening of membrane. Thickening of membrane is due to increased production or glycation of matrix proteins due to hyperglycemia. [437]

Our results of microalbuminuria in diabetic patients are in accordance with study done by Sheng et al.[435] Microalbuminuria is a significant risk factor for cardiovascular mortality in type 2 diabetes.[464] Diabetes exerts its effects on glomerular permeability in the initial stages of diabetic nephropathy, i.e. at or before the appearance of microalbuminuria. The increased transglomerular passage of albumin is the major source of microalbuminuria.[465] The other hypotheses is failure of tubular reuptake of albumin. These early changes establish the milieu in which more advanced changes of overt diabetic nephropathy develop.

The presence of microalbuminuria is a predictor of worse outcomes for both renal and cardiac patients. Microalbuminuria does not directly cause cardiovascular events; it serves as a marker for identifying those who may be at increased risk. Microalbuminuria is caused by glomerular capillary injury and so may be a marker for diffuse endothelial dysfunction. According to Steno hypothesis, albuminuria might reflect a general vascular dysfunction and leakage of albumin and other plasma macromolecules such as low density lipoproteins into the vessel wall that may lead to inflammatory responses and in turn start the atherosclerotic process.[464, 466]

Early diabetes is characterized by increased microvascular pressure and flow, which act as injurious stimuli to the endothelium (haemodynamic hypothesis), leading to excess matrix protein production, capillary basement membrane thickening, sclerosis and impaired vasodilatory function. In addition, the Steno hypothesis proposes that a high risk of microangiopathy, as well as a tendency to large vessel disease, is a function of increased permeability of the vascular endothelial cell layer.[20] This increase in permeability has been attributed to changes in cell surface electrostatic charge. Jensen concluded that the presence of clinical atherosclerotic cardiovascular disease is associated with both renal and systemic transvascular leakiness for albumin. The 'response-to-injury' hypothesis proposes that 'injury' to the endothelium, e.g. by haemodynamic strain or dyslipidaemia, is the initiating event in the atherogenesis.[468] One type of 'response' to such 'injury' is increased transendothelial permeability of macromolecules. This systemic leakiness includes other macromolecules such as lipoproteins, thereby perhaps augmenting their atherogenic effect. Transendothelial passage of albumin and low-density lipoprotein (LDL) has been correlated to subsequent atherogenesis.[469]

Both acute hyperglycaemia and glycaemic control, as measured by HbA1c were also found to be significant predictors of endothelial dysfunction. Furthermore, hyperglycaemia induces protein-C activation and depletion of nitric oxide synthase co-factors, thereby reducing nitric oxide bioavailability. Advanced glycation end products (AGEs) directly reduce the nitric oxide that is produced by the endothelial cell.

The results of microalbumin of Group III are similar to results obtained by Korah et al in type 2 diabetes with complications.[437] In our study we found significant association of microalbuminuria with diabetic complications. Manaviat et al [439] also showed association between microalbuminuria and retinopathy in type 2 diabetes and Kundu et al showed association of microalbuminuria with nephropathy in type 2 diabetes.[309]

Conclusion:

The decreased levels of serum nitric oxide in all the study groups are suggestive of endothelial dysfunction which begins at metabolic syndrome phase. Finding of our study suggest that insulin resistance, dyslipidemia and hyperglycemia reduces the bio availability of nitric oxide.

7.6 Genetic analysis

Nitric oxide (NO) plays fundamental role in the regulation of endothelial function and vascular tone in many organs including kidney. It also inhibits platelet aggregation, leukocyte adhesion to vascular endothelium & oxidation of LDL [471] NO is produced by NOS enzyme which exhibits three isoforms: neuronal (nNOS), inducible (iNOS) and endothelial (eNOS). Endothelium-derived NO is synthesized from L-arginine by NO synthase coded by eNOS (eNOS or NOS3) gene, mapped on chromosome 7q36. Upon release, NO diffuses rapidly through the cell membrane and relaxes neighboring vascular smooth cells through the production of guanine 3'5'-cyclic monophosphate (cGMP). Impairment of NO production causes endothelial dysfunction contributing to the development of insulin resistance, type 2 diabetes, chronic renal failure and cardiovascular complications including hypertension and hypercholesterolemia.[472] Due to the importance of eNOS in the generation of NO that regulates endothelial-dependent vasodilation in many organs, several epidemiological studies have evaluated the role of eNOS gene polymorphisms in the eNOS enzyme activity and NO availability in type 2 diabetes, cardiovascular and renal diseases. Several polymorphisms of the eNOS gene that have been identified are:

1. 786T>*C* polymorphisms in the promoter region

A single nucleotide polymorphism (SNP), -786T>C point mutation, was identified in the 5`-flanking region of the eNOS gene involving a substitution of thymine (T) to cytosine (C) at a locus 786bp upstream of eNOS gene.[472,473] This results in a significant reduction in the eNOS gene promoter activity.[474] This polymorphism has the potential to influence mRNA transcription,[475] where it is associated with less placental mRNA and lower serum nitrite/nitrate concentrations.[476] One manifestation of the T-786>C mutation is increased risk for coronary spasm and coronary artery diseases.[477]

2. Variable nucleotide tandem repeat (27 VNTR) of 4a/4b polymorphism in intron 4

A 27-bp repeat polymorphism in intron 4 of the eNOS gene, has been associated with variations in plasma levels of nitrite and nitrate (NOx)-stable metabolites of NO.[474] The intron 4 (4a/4b) VNTR polymorphisms of the gene encoding eNOS has been shown to segregate with lower plasma NO metabolites in non pregnant Japanese, [471,478-480] Also this polymorphism was found to be associated with recurrent pregnancy loss (RPL) in Caucasians. [472, 479,481] The heterozygosity for 4a allele and 4b allele of eNOS (4a/4b) was found to be associated with RPL in Austrian women. It is speculated that carriers of 4a/4b and subsequently reduced NO serum levels are at increased risk for impaired placental perfusion and infarction.[480]

3. Glu298Asp or 894G>T polymorphism in exon 7

The Glu298Asp missense mutation encoded by exon 7 of the eNOS gene is a common variant of eNOS that has a guanine (G) to thymine (T) transversion at nucleotide position 894, resulting in a replacement of glutamic acid by aspartic acid at codon 298 (Glu298Asp). 894G>T is associated with reduced basal NO production and has been reported to be associated with an increased risk for myocardial infarction, coronary artery spasm, hypertension, placental abruption and preeclampsia but not with recurrent pregnancy loss (RPL).[474 ,477,479,481]

The Glu298Asp polymorphism causes a structural change of the eNOS protein and reduces eNOS activity.Therefore, the gene coding for eNOS can be a good candidate gene for evaluating risk of occurrence of diseases in metabolic syndrome, type 2 diabetes with and without complications and its association with endothelial markers and IR (ENOS in MS) and *hence the SNP in this gene was included in our study.*

7.6.1 Flowchart of genetic analysis

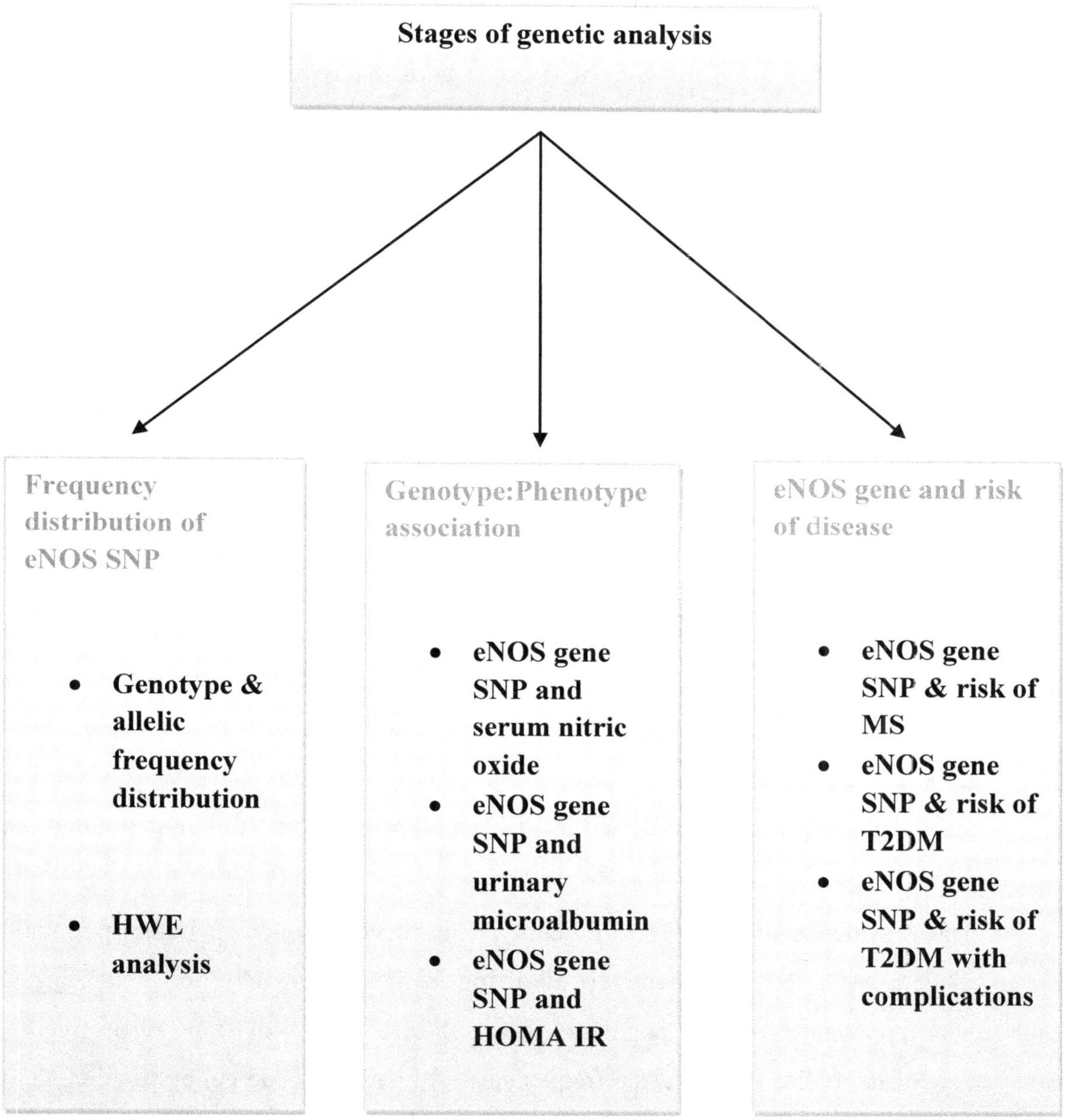

7.6.2 Genotype Frequency

i) Genotype frequency in Control Group

RFLP analysis of eNOS gene G894T showed genotype frequency as GG, GT, TT. The frequency distribution in control Group was 84%, 16%, 0% respectively as shown in table 6.5a. This was in accordance with Hardy-Weinberg equilibrium. ($\chi 2$ test = 2.26, degree of freedom=1, P>0.05 N.S).

Angeline et al [5] in their study reported 71.62% and 29.37% frequency for GG and GT genotype in healthy controls in south Indian population and found that the homozygous TT genotype was absent.

Shankarishan et al determined the prevalence of eNOS gene variant among both the tea garden sect and Indigenous community of Assam. The frequencies of the GG, GT and TT genotypes observed among the Tea garden community were 77.0%, 21.0% and 2.0% per cent respectively and those among the Indigenous Assamese community were 65.0%, 32.0% and 3.0% respectively. [482]

In a study about prevalence of eNOS Glu298Asp polymorphism in healthy volunteers from Northern India region, Srivastava et al reported the distribution of GG, GT and TT to be 71.22 %, 28.06% and 0.72%, respectively. [483] Tripathi et al , reported 74.5%, 24.2% and 1.3 % frequency for GG, GT and TT genotypes in the north Indian population. [484]

The prevalence of GG, GT, TT genotype frequencies found in Italian population were 44.2%, 49%, 6.8% respectively showing highest frequency for heterozygous (GT) genotype whereas in UK population it was 47.8%, 42% and 10.2% respectively. Among the Asian studies reported, the frequencies of the eNOS gene genotypes GG, GT and TT were 90.95%, 8.83% and 0.2% respectively among Chinese population and 86.7%, 13.2 % and 0.2% respectively in Japanese. [290]

The frequencies of these genotypes observed in our healthy control population is in agreement with the genotype frequencies reported by Angeline et al and Shankarishan et al [5,482] **Since the frequency of GG genotype is higher in healthy control it can be said that this may be a protective genotype against endothelial damage.**

ii) Genotype frequency in Study Groups

In our study, the frequency of the GG, GT and TT genotypes of eNOS gene in Group II (MS) was found to be 65%, 27% and 8% respectively. It was 53%, 30% and 17% respectively, in Group III (T2DM) and 42%, 35% and 23% respectively in Group IV (T2DM with complications, refer to table 6.5a)

The prevalence of heterozygous (GT) and homozygous TT genotype for eNOS - G894T SNP was found to be higher in all study groups as compared to controls. The presence of high frequency of eNOS gene genotype GT and TT in study group may be indicative of occurrence of the disease.

In a study done by Liu et al on association of eNOS gene polymorphism with metabolic risk factors, the genotype frequencies of GG, GT and TT genotype was found to be 77.6%, 22.1% and 0.3% respectively. The frequency of GT genotype for eNOS gene in metabolic syndrome patients reported by us is in agreement with Liu et al. [485]

Ianas et al reported the genotype frequency for eNOS gene in MS as 32.43% (GG), 51.35% (GT) and 16.2 %(TT) [294] while Monti et al reported 32.7%, 39.6% and 27.7% respectively for GG, GT and TT in MS. [287]

Shah et al examined association of eNOS gene polymorphism in type 2 diabetic patients (T2DM) and reported the frequencies of GG, GT and TT genotypes as 44%, 46% and 10% respectively.[305] In another study, Angeline et al reported 53.07%, 39.23% and 7.69% frequencies for GG, GT and TT genotype in diabetic population. [5] Shoukry et al observed 49.5%, 39.55% and 12% frequencies of GG, GT and TT genotypes in type 2 diabetes patients.[304]

Shah et al in their study of eNOS SNP in diabetic complications reported the frequencies of GG, GT and TT genotypes as 47%, 43% and 10 % respectively in diabetic nephropathy, [305] whereas Shoukry et al observed it to be as 33% ,47%, 20% respectively in these patients. [304]

Our finding of high prevalence of GT and TT genotype in type 2 diabetes and type 2 diabetes with complications is in accordance with studies done by Shah et al, Shoukry et al and Angeline et al.

iii) Allelic frequency in Control Group

In the present study the allelic frequency of G and T allele in control Group was 92% and 8 % resp. Nishevitha et al[290] demonstrated 87% and 13% frequency of G and T allele respectively in south Indian Tamil healthy population whereas Angeline et al observed 85 % and 15% respectively.

Shankarishan et al reported G and T allele frequency among the Tea garden community of Assam as 87% and 13% and among the indigenous Assamese community as 81 % and 19% respectively. [482] In the study by Srivastava et al the frequency of G allele was 85 % and T allele was 14% in healthy controls. Our frequency of G and T allele is similar to the frequencies quoted by the studies. [483]

iv) Allelic frequency in Study Group

The frequency of G and T allele was found to be 78%, 22% in Group II, 68%, 32% in Group III, and 59%, 41 % in Group IV resp. and is presented in table 6.5b.

The frequencies for the G & T alleles in patients with metabolic syndrome reported by Ianas et al[294] was 58% and 41.9 % respectively, while Liu et al observed as 83% and 13 % for the same. The allelic frequency reported by our study is not in accordance to those reported. [485]

In case of type 2 diabetes, the frequencies of G and T allele of eNOS G298T was found to be 52% and 48% respectively by Angeline et al,[5] With regards to patients of diabetics nephropathy, Shourky et al reported 56.5% and 43.5% respectively as G and T allelic frequency, while Shah et al [305] reported 68% and 32% respectively. Our finding of G and

T allele frequency in patients with and without complications are in accordance with studies reported by Shah et al, Angeline et al and Shoukry et al [5,305,304]

7.6.3 Risk of occurrence of diseases

Fischer's exact test and 2x2 contigency table (for odds ratio) were employed as the most suitable statistical tool for genetic analysis. GT+TT genotypes were combined for genotype: phenotype analysis due to low frequency of TT in our population. Odds ratio of study groups is presented in table 6.5c.

The frequency of GG and (GT+TT) genotype of control (Group I) and metabolic syndrome (Group II) when compared by Chi -square analysis showed an odds ratio of 2.82 for metabolic syndrome indicating high chance for occurrence of metabolic syndrome among the healthy population ($p<0.05$).

Ianas et al investigated the association of eNOS G894T polymorphism and endocrine-metabolic changes in MS and found significant association of this polymorphism with MS with an odds ratio for GT =2.09 and for TT= 3.08.[294]

The genotype frequency of patients of type 2 diabetes and type 2 diabetes with complications, exhibited an odds ratio of 4.74 and 7.06 respectively, again indicating strong risk for occurrence of diabetes mellitus ($p<0.001$) and diabetes with complications ($p<0.000$) among the selected population. Refer table 6.5c.

Angeline et al in a study about association of G894T SNP with type 2 diabetes mellitus in South Indian (Tamil) population found that (GT+TT) is associated with diabetes mellitus with odds ratio of 5.28. [5]

7.6.4 Genotype: Phenotype Association

For studying contribution of eNOS gene polymorphism on the function of endothelium, the following phenotypic characteristics were considered viz:

- Serum Nitric oxide
- Urinary microalbumin
- HOMA IR

Insulin regulates nitric oxide production in endothelial cells by subsequently activating the intracellular enzymes PI3K (phosphoinositide 3-kinase) and Akt [486] which phosphorylates and in turn activates eNOS. The association of serum nitric oxide levels with Glu298Asp was done to study functional effect of this polymorphism on nitric oxide synthesis. Urinary microalbumin level is also related to eNOS gene polymorphism as NO plays important role in control of renal and glomerular hemodynamics.[470] The association of eNOS gene SNP to insulin resistance was assessed using HOMA IR.

For the association studies we performed:

- Within group analysis- in which the association of each genotype with particular phenotype characteristic was considered within each group and presented in table 6.5e.
- Intergroup analysis – in which the genotype of each group was compared with the same genotype of other groups as follows: (represented in table 6.5f)
    - Healthy controls (Group I) with subjects of MS (Group II)
    - Metabolic syndrome subjects (Group II) with type 2 diabetes patients (Group III)
    - Diabetic patients (Group III) with patients of diabetic complications (Group IV)

The purpose of such specific group comparison is keeping in mind our hypothesis that MS is one end of the continuum of the disease of diabetes with the angiopathic complications on the other end.

i) Association with Serum nitric oxide levels in different study groups

Serum Nitric oxide levels were considered for within group and between group analysis for studying its effect on genotypes.

- **Within Group analysis:**

In this study we noted that metabolic syndrome subjects with GG genotype had higher NO levels than those of (GT+TT). (Refer to table no 6.5d) and thus are probably protected from endothelial damage.

Our observation that GG genotype is a protective factor is based on the presence of higher frequency of this genotype among control. Accordingly, serum nitric oxide level of subjects with this genotype in all the groups is expected to be higher as compared to those of (GT+TT) genotype. Such an observation was seen only in the metabolic syndrome group (0.29 umols/l vs 0.22 umols/l, $p<0.02$, refer to table 6.5d & 6.5e). This may be because MS is the initial stage of onset of full blown disease of diabetes, when the association of genotype is visible. Therefore, the damage to endothelium begins at early stage and once done, continues to remain further throughout the disease increasing in severity in patients of diabetes with complications. The association of genotype to NO is not observed in type 2 diabetes and diabetes with complications as many interplaying factors contribute to endothelial damage in addition to genotype variants.

When serum NO levels of GG genotype subjects were compared with that of (GT+TT) of both type 2 diabetes patients as well as those with diabetic complications, no significant difference was observed. This can be explained on the basis of our hypothesis that metabolic syndrome is the prediabetic phase of overt diabetes. Hence in Group III, at the time of estimation of serum nitric oxide levels, these patients had already crossed the preliminary phase of MS when damage begins.

Since in the present study, serum NO levels were estimated in type 2 diabetes diagnosed patients (Group III) who probably had already crossed the stage of MS when the damage begins. Hence the protective effect of GG genotype could not be directly observed. These patients might have had better serum nitric oxide levels in their metabolic syndrome phase.

- **Intergroup analysis for serum nitric oxide**

Group I (Control) with Group II (Metabolic syndrome)

In our study we found that serum nitric oxide levels of subjects with GT+TT genotype of metabolic syndrome group showed significant decrease when compared to the same genotype subjects of control ($p<0.001$), while serum NO levels of GG genotype subjects did not differ between these two groups, thus proving that GG may be a protective genotype contributing to higher production of nitric oxide.

Group II (Metabolic syndrome) with Group III (Type 2 diabetes)

The genotype analysis between the groups of metabolic syndrome & type 2 diabetes showed that the GG and (GT+TT) subjects of type 2 diabetes have significantly lower NO levels as compared to those of MS. ($p<0.001$ and $p<0.007$ respectively).

Group III (Type 2 diabetics) with Group IV (Diabetes with complications)

When the serum nitric oxide levels of both the GG and (GT+TT) genotypes of type 2 diabetes patients and those with complications were compared, no significant difference was observed.

Complications in diabetes are due to microangiopathies attributed to endothelial damage which in turn affects the production of the protective NO. Although eNOS gene SNP contributes to NO production per se,' it may not be the only responsible causative factor. The role of other contributing agents such as gluctoxicity, AGE and Apo B, highly active polyol pathway as well as role of inflammation can not be undermined.

ii) Association of Microalbuminuria

Urinary microalbumin, the second endothelial marker of GG and (GT +TT) genotypes subjects in all study groups as well as healthy control showed no statistical significance.

- **Within group analysis**

Urinary microalbumin of GG genotype subjects of all study groups when compared with those of (GT +TT) genotype of individual group, no statistically significant relationship was observed.

- **Inter group analysis for microalbuminuria**

When the urinary micoalbumin values of both GG and (GT+TT) subjects of both MS as well as Control group were compared, no significant difference was observed. The urinary Microalbumin levels of GG genotype of type 2 diabetic patients (Group III) when compared with the same genotype of diabetic patients showing complications(Group IV) significant increase was observed ($p<0.001$). Similarly microalbumin of patients with (GT+TT) genotype between Group III & IV also showed significant increase ($p<0.001$).

Thus it can be stated that variance of G894T has no association with urinary microalbumin in any of the conditions. Although increased urinary microalbumin is seen in type 2 diabetes and its complication, this finding is independent of genotype variants.

iii) Association of Genotype with HOMA IR

HOMA IR values of subjects with different genotypes were taken into consideration for within and inter group comparison with the genotypes.

- **Within Group analysis**

HOMA IR an accepted indicator of insulin resistance was measured and correlated with eNOS gene polymorphism. High values of HOMA IR were observed in (GT+TT) genotype of MS subjects when compared with (GG) genotype in the same group (6.47 vs 5.97, $p<0.05$, refer to table 6.5d & 6.5e). This may be because insulin resistance impairs the production of NO and favors the production of endothelin-1, the vasoconstrictor which have effect on isolated arterioles.[487] The comparison of HOMA IR values of GG genotype subjects with that of (GT +TT) in both type 2 diabetes with and without complications showed no significant difference within each of the groups. Although hyperinsulinemia is seen in diabetic patients with and without complications, the insulin utilization in these patients is affected. Also there is accompanied hyperglycemia.

Therefore the values for HOMA IR remain unaffected in these patients due to genotype variants.

- **Inter Group analysis:**

Group I (Control) with Group II (Metabolic syndrome)

In our study HOMA IR values of metabolic syndrome subjects with GG & GT+TT genotype showed significant increase when compared to the same genotype subjects of control ($p<0.001$), thus confirming the finding of persistent insulin resistance in metabolic syndrome.

Group II (Metabolic syndrome) with Group III (Type 2 diabetes)

The genotype analysis between the groups of metabolic syndrome & type 2 diabetes showed that both GG and (GT+TT) subjects of type 2 diabetes had significantly higher HOMA IR as compared to MS ($p<0.01$ and $p<0.009$ respectively). In frank diabetes, during the phase of hyperglycemia, the β-cells of pancreatic islets of Langerhans secrete more insulin to promote utilization of glucose, which explains the finding of higher values for both insulin and glucose with increased values for HOMA IR in type 2 diabetes irrespective of genotypes.

Group III (Type 2 diabetes) with Group IV (Diabetes with complications)

The HOMA IR of both the GG and (GT+TT) genotype patients of type 2 diabetes as well as those with diabetic complications showed no significant difference indicating that the genotype variants have no effect on HOMA IR. Diabetic complications are the phase characterized by glocotoxicity in which the secreted insulin fails to produce necessary physiologic effect resulting in insulin resisitance. Therefore although insulin resistance is observed in type 2 diabetes and its complications, it may not be related to eNOS gene polymorphism.

In conclusion, the effect of eNOS gene G894T SNP is observed only in MS group in terms of changes in phenotypic variables such as HOMA IR and Serum nitric oxide while no association was observed for urinary microalbumin. The genotype variants seem to influence the endothelial damage as seen by the changes in NO levels. The genotype also influences insulin resistance, the hallmark of MS as seen by changes in HOMA IR in these subjects.

It is essential to identify subjects with MS at an early stage so as to counsel them about life style modification. Studying the eNOS G894T in these subjects helps to predict their predisposition to develop type 2 diabetes and cardiovascular diseases.

7.6.5 Conclusion of genetic analysis

The following observations were made:

1. GG is predominant genotype and T is minor allele.
2. The presence of GG might be a protective factor for endothelium.
3. Presence of GT &TT might be predispose a subject to occurrence of MS,T2DM and its complications. (Odds ratio for metabolic syndrome, Type 2 Diabetes and Diabetic complications was observed to be 2.82, 4.74 and 7.06 respectively).
4. eNOS gene polymorphism is correlated with both nitric oxide production and insulin resistance in metabolic syndrome.
5. Genotype may not be responsible for progression of disease from metabolic syndrome to type 2 diabetes as no correlation of any particular genotype was found with type 2 diabetes or diabetic complications.

Chapter 8

Summary and Conclusion

Summary and conclusion

Metabolic syndrome is characterized by clustering of risk factors, which predisposes subjects to increased risk of diabetes and cardiovascular disease (CVD). Insulin resistance is considered as the hallmark of metabolic syndrome which is a pre diabetic phase. Although type 2 diabetes is a heterogeneous disease, most patients with T2DM have insulin resistance and the MS before onset of type 2 diabetes. In fact, insulin resistance, hyperinsulinemia, dyslipidemia, and obesity precede the progression to type 2 diabetes in 75 to 85% of patients.

In our study population we measured insulin resistance by HOMA IR and correlated with factors of metabolic syndrome like WHR, BMI in all the study groups using univariate and multivariate regression analysis. Both these parameters showed good degree of correlation with HOMA IR. The estimation of lipid profile reveled that dyslipidemia is a typical feature of metabolic syndrome and is also observed in type 2 diabetes and its complications further affecting health of endothelium. The indices of insulin resistance and dylipidemia like TG/HDL-C ratio, Apo B were correlated with HOMA IR and a good degree of correlation was observed. Hypertension was observed in all the study groups supporting the fact that the hypertension is associated with insulin resistance, one of the pathophysiologic features underlying metabolic syndrome. The health of endothelium was assessed by measuring serum nitric oxide and urinary microalbumin levels. The endothelial markers were also correlated with HOMA IR to assess the damage and effect of insulin resistance on endothelium. It was observed that damage to endothelium is initiated at early stage of metabolic syndrome during which phase though damage is not presented clinically, it becomes significant in presence of IFG, IGT and/or altered lipid profile.

In diabetic population (with and without) a disturbed glucose homeostasis was observed in the form of hyperglycemia, raised HbA1c and elevated HOMA IR. The lipid markers and endothelial markers were also alerted.

To study the contribution of genetic factors, SNP analysis was done for Glu298Asp of eNOS gene located on chromosome 7. The genotype and allelic frequency was determined. The risk of occurrence of disease was studied. The observations of genetic analysis were as follows:

1. GG is predominant genotype. G is a dominant allele and T is minor allele
2. The presence of GG might be protective factor for endothelium as the levels of serum nitric oxide in patients with GG genotype was more.
3. Presence of GT &TT might be predisposing a subject to occurrence of MS, T2DM and its complications. The odds ratio for MS, Type 2 Diabetes and Diabetic complications was observed to be 2.82, 4.74 and 7.06 respectively.
4. eNOS gene polymorphism is correlated with both nitric oxide production and insulin resistance in metabolic syndrome.
5. Genotype may not be responsible for progression of disease as no correlation of any particular genotype was found with type 2 diabetes or diabetic complications.

Drawing conclusion from our study we confirm that metabolic syndrome the pre diabetic, silent phase is at one end of the continuum of metabolic disease with full blown type 2 diabetes at other end. The duration taken by a metabolic syndrome individual to develop frank diabetes may depend on genotype of eNOS gene. It can be seen from our study that GG is a protective genotype and individuals with GT and TT genotype are at higher risk for development of type2 diabetes, followed by endothelial dysfunction.

In order to prolong this progression from metabolic syndrome to type 2 diabetes it is highly recommended to identify individuals at the metabolic syndrome stage to provide better quality of life at their later age which will reduce the morbidity and mortality rate in India due diabetic complications like CVD.

India is known to be the second diabetic capital of world. It is well documented that metabolic syndrome is a prediabetic stage. Urban lifestyle and accompanying stressors contribute to faster progression of metabolic syndrome to diabetic stage. Therefore to reduce the healthcare economic burden of nation, it is important to identify individuals at metabolic syndrome phase and take necessary measures.so as to prevent their progress in to diabetic stage.

Model showing predisposition and prevention for disease

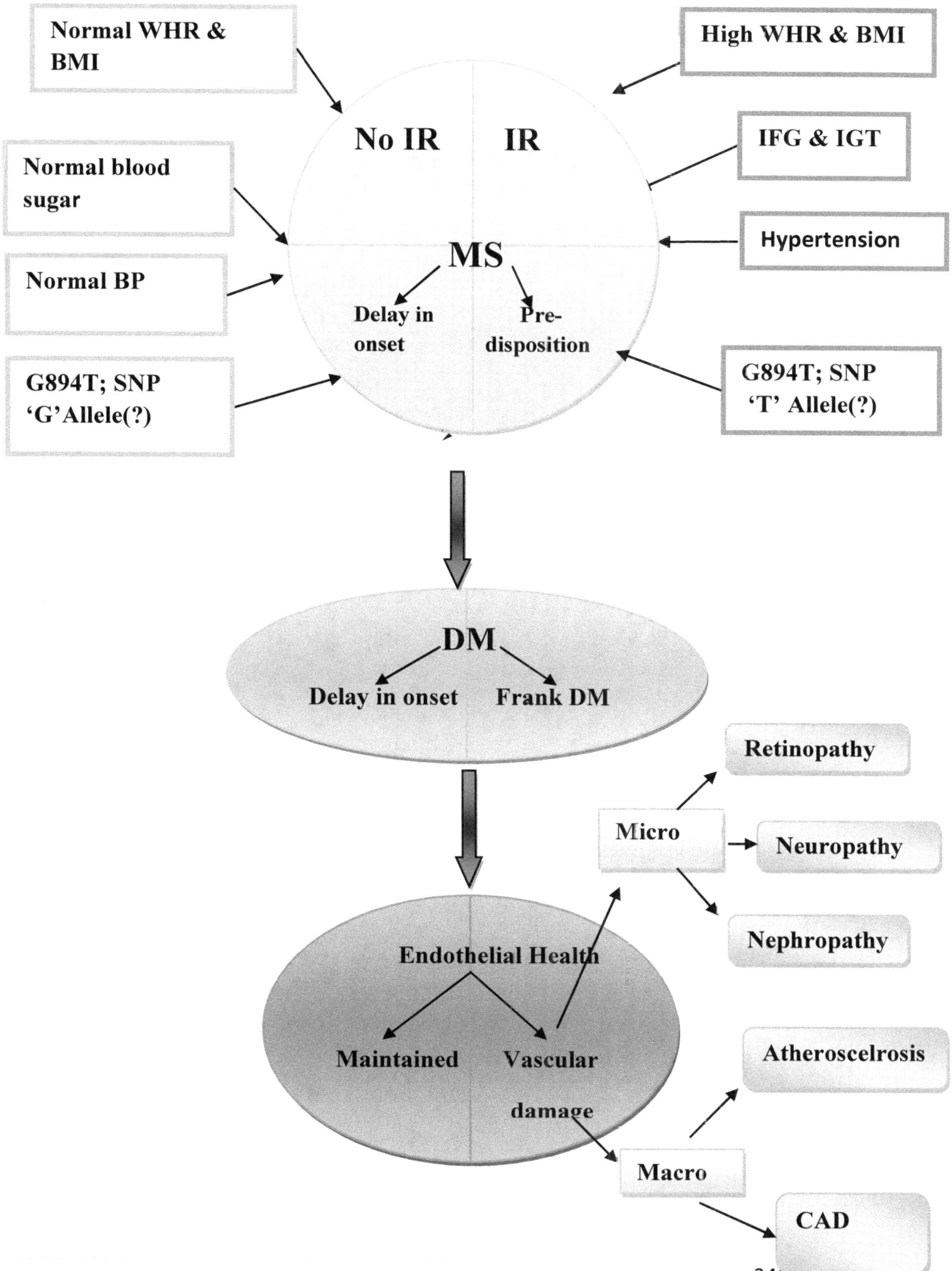

Bibilogrpahy

Bibilography:

1. Yusuf S, Ounpuu S. Tackling the growing epidemic of cardiovascular disease in South Asia. J Am Coll Cardiol 2001; 38:788-9.

2. International Diabetic federation. The Diabetic atlas; 5th edition. http://www.idf.org/diabetesatlas cited on 2nd February 2012

3. Mohan V. Why are Indians more prone diabetes? J Assoc physicians India 2004;52:468-474.

4. Petersen KF, Shulman GI. Etiology of insulin resistance. Am J Med 2006;119: S10-S16.

5. Angeline T, Krithiga HR, Isabel W, et al. Endothelial nitric oxide synthase gene polymorphism (G849T) & diabetes mellitus (type 2) among south Indians. Oxidative medicine and cellular longetivity 2011:1-2.

6. Shankar P, Sundarka M. Metabolic syndrome: Its pathogenesis & management. JIACM 2003;4(4):275-281.

7. American Diabetes Association. Consensus Development Conference on Insulin Resistance.Diabetes Care 1998; 21:310–314.

8. Albert KG, Zimmet PZ. Definition, diagnosis and classification of diabetes mellitus provisional report of WHO Consultation. Diabet Med 1998; 15:539–53.

9. Expert panel on detection, evaluation & treatment of high blood cholesterol in adults. Executive summary of the Third Report of The National Cholesterol Education Program (NCEP) Expert Panel on Detection, Evaluation, and Treatment of High Blood Cholesterol in Adults (Adult Treatment Panel III). JAMA 2001; 285:2486–2497.

10. Hunt KJ, Haffner RG, Ster MP. National cholesterol education program versus world health orgnisation metabolic syndrome in relation to all cause & cardiovascular mortality in the San Antonio Heart Study. Circulation 2004;110:10:1251-1277.

11. Cozma A, Orasan O, Sampelen D, Fodor A, et al. Endothelial dysfunction in metabolic syndrome. Rom J Inter Med 2009; 47:2:133-140.

12. Powers AC. Diabetes meliitus, Chapter 344. Harison's principle of Internal Medicine;18th edition McGraw Hill 2012:2968-2996

13. Reports of the Expert Committee on the Diagnosis and Classification of Diabetes. Diabetes Care 1997; 20:1183–1197.

14.Wild S, Sicree R, Roglic G, King H, Green A. Global Prevalence of Diabetes Estimates for the year 2000 and projections for 2030. Diabetes Care 2004; 27:1047-1053.

15. Chan JC, Malik V, Jia W, et al. Diabetes in Asia: Epidemiology, risk factors and pathophysiology. JAMA 2009; 301:2129–2140.

16. Ramachandran A, Wan Ma RC, Snehalatha C. Diabetes in Asia. Lancet 2010;375:408–418.

17. IDF Diabetes Atlas, 6th edition. International Diabetes Federation. 2013.

18. Yang W, Lu J, Weng J, et al. Prevalence of diabetes among men and women in China. N Engl J Med 2010; 362:1090–10101.

19. Mohan V, Pradeepa R. Epidemiology of diabetes in different regions of India. Health Administrator 2009; 22:1–18.

20. Ramachandran A, Snehalatha C, Kapur A, et al. High prevalence of diabetes and impaired glucose tolerance in India: National Urban Diabetes Survey. Diabetologia 2001; 44:1094–1101.

21. Sicree R, Shaw J, Zimmet P. Diabetes and impaired glucose tolerance. Diabetes Atlas. International Diabetes Federation. 3rd ed. Belgium: International Diabetes Federation; 2006:15-103.

22. Ahuja MMS. Epidemiological studies on diabetes mellitus in India. Epidemiology of diabetes in developing countries. New Delhi: Interprint. 1979:29-38.

23. Sridhar GR, Rao PV, Ahuja MMS. Epidemiology of diabetes and its complications. In: RSSDI textbook of diabetes mellitus. Research Society for the Study of Diabetes in India 2002:95-112.

24. Ramachandran A, Snehalatha C, Dharmaraj D, Viswanathan M. Prevalence of glucose intolerance in Asian Indians. Urban-rural difference and significance of upper body adiposity. Diabetes Care 1992; 15:1348-55.

25. Ramachandran A, Snehalatha C, Latha E, Vijay V, Viswanathan M. Rising prevalence of NIDDM in an urban population in India. Diabetologia 1997; 40:232-237.

26. Shah SK, Saikia M, Burman NN, Snehalatha C,Ramachandran A. High prevalence of type 2 diabetes in urban population in north eastern India. Int J Diabetes Dev Countries 1999; 19:144-147.

27. Zargar AH, Khan AK, Masoodi SR, Laway BA, Wani AI, Bashir MI, et al. Prevalence of type 2 diabetes mellitus and impaired glucose tolerance in the Kashmir Valley of the Indian subcontinent. Diabetes Res Clin Pract 2000; 47:135-136.

28. Alberti KG, Zimmet PZ. Definition diagnosis and classification of diabetes mellitus and its complications. Part I: Diagnosis and classification of diabetes mellitus, provisional report of a WHO consultation. Diabetes Med 1998; 15:539-553.

29. Gupta A, Gupta R, Sarna M, Rastogi S, Gupta VP, Kothari K. Prevalence of diabetes, impaired fasting glucose and insulin resistance syndrome in an urban Indian population. Diabetes Res Clin Pract 2003; 61:69-76.

30. Deo SS, Zantye A, Mokal R, Mithbawkar S, Rane S, Thakur K. To identify the risk factors for high prevalence of diabetes and impaired glucose tolerance in Indian rural population. Int J Diab Dev Ctries 2006; 26:19-23.

31. Menon VU, Kumar KV, Gilchrist A, Sugathan TN, Sundaram KR, Nair V, et al. Prevalence of known and undetected diabetes and associated risk factors in central Kerala - ADEPS. Diabetes Res Clin Pract 2006; 74:289-94.

32. Mohan V, Deepa M, Deepa R, Shantirani CS, Farooq S,Ganesan A, et al. Secular trends in the prevalence of diabetes and glucose tolerance in urban South India - the Chennai Urban Rural Epidemiology Study (CURES-17). Diabetologia 2006; 49:1175-1178.

33. Raman KV, Joseph A, Soman CR. High prevalence of type 2 diabetes in an urban settlement in Kerala, India. Ethn Health 1999; 4:231-9.

34.Joshi SR, Saboo B, Vadivale M, Dani SI, et al. Prevalence of Diagnosed and Undiagnosed Diabetes and Hypertension in India-Results from the Screening India's Twin Epidemic (SITE) Study. Diabetes Technology & Therapeutics 2012; 14(1):8-15.

35. Anjana RM, Pradeepa R, Deepa M, Datta M,Unnikrishnana R,Mohan V, et al. ICMR India Collabrative study group. Diabetologia 2011;54(12):3022-3027.

36. Qiao Q, Hu G, Tuomilehto J, et al. Age- and sex-specific prevalence of diabetes and impaired glucose regulation in 11 Asian cohorts. Diabetes Care 2003; 26(6):1770-1780.

37. Eknoyan G, Nagy J. A History of Diabetes Mellitus or How a Disease of the Kidneys Evolved Into a Kidney Disease. Advances in Chronic Kidney Disease 2005; 12(2):223-229.

38. http://www.diabetesindia.com/diabetes/history_diabetes3.htm accessed on 9th sep 2013.

39. **http://www.defeatdiabetes.org** accessed on 9th sep 2013.

40. DeFronzo R. From the triumvirate to the ominous octet: a new paradigm for the treatment of type diabetes mellitus. Diabetes 2009; 58:773–795.

41. Boden G, Chen X, Stein TP. Gluconeogenesis in moderately and severely hyperglycemic patients with type 2 diabetes mellitus. Am J Physiol 2001; 280: E23 – E30.

42. Faerch K, Vaag A, Holst JJ, Hansen T, Jørgensen, T, Borch-Johnson K. Natural history of insulin sensitivity and insulin secretion in the progression from normal glucose tolerance to impaired fasting glycemia and impaired glucose tolerance: the Inter99 Study. Diabetes Care 2009; 32:439–444.

43. Lorenzo C, Wagenknecht L, D'Agostino R, Rewers M, Karter A, Insulin resistance,β-cell dysfunction and conversion to type 2 diabetes in a multiethnic population. Diabetes Care 2010; 33:67–72.

44. Kendall DM, Cuddihy RM, Bergenstal RM. Clinical application of incretin-based therapy: therapeutic potential, patient selection, and clinical use. Am J Med 2009; 122:S37–S50.

45. Zeggini E, Scott LJ, Saxena R, Voight BF, Marchini JL, Hu T, et al. Meta-analysis of genome-wide association data and large-scale replication identifies additional susceptibility loci for type 2 diabetes. Nat Genet 2008;40: 638– 645.

46. Voight BF, Scott LJ, Steinthorsdottir V, Morris AP, Dina C, Welch RP. et al. Twelve type 2 diabetes susceptiblility loci identified through large scale association analysis. Nat Genet 2010; 42:579 – 589.

47. Kendall DM, Cuddihy RM, Bergenstal RM, Clinical Application of Incretin-Based Therapy: Therapeutic Potential, Patient Selection and Clinical Use Am J Med 2009;122(6A):S37-S44.

48. Kahn SE.The Importance of β-Cell Failure in the Development and Progression of Type 2 Diabetes. The Journal of Clinical Endocrinology & Metabolism 2001; 86(9):4047–4058.

49. DeFronzo RA. The triumvirate: beta cell, muscle, liver – a conclusion responsible for NIDDM. Diabetes 1988;37(6):667-684.

50. Gerich JE. Control of glycemia. Baillier's Best Practice and Research in Clinical Endocrinology & Metabolism 1993;7(3):551-586.

51. Wilcox G. Insulin and Insulin Resistance. Clin Biochem Rev 2005; 26:19-21.

52. Rosenfeld L. Insulin: Discovery and Controversy. Clin Chem 2002;48:12 2270–2288.

53. Unnikrishnan AG, Kalra S, Baruah M. The other insulin story of 1921. Indian J Endocrinol Metab 2011;15(3):147–148.

54. Keen H. International Textbook of Diabetes Mellitus (2nd ed) John Wiley & Sons, New York; 1997:899-928.

55. Dodson G, Steiner D. The role of assembly in insulin's biosynthesis. Curr Opin Struct Biol 1998;8:189-94.

56. Porksen N, Hollingdal M, Juhl C, Butler P, Veldhuis JD, Schmitz O. Pulsatile insulin secretion: detection, regulation, and role in diabetes. Diabetes 2002; 51(1):S245-54.

57. Bratanova-Tochkova TK, Cheng H, Daniel S, et al. Triggering and augmentation mechanisms, granule pools and biphasic insulin secretion. Diabetes 2002; 51(1):S83-90.

58. Nielsen JH, Galsgaard ED, Moldrup A, et al. Regulation of beta-cell mass by hormones and growth factors Diabetes 2001;50(1):S25-9.

59. De Lonlay, Saudubray J-M. Persistent hyperinsulinaemic hypoglycaemia. In: Fernandes J, Sudubray J-M, van den Berghe editors Inborn Metabolic Diseases: Diagnosis and treatment. (3rd ed): Springer, Heidelberg Germany; 2000:117-26.

60. Soria B, Quesada I, Ropero AB, Pertusa JA, Martin F,Nadal A. Novel players in pancreatic islet signaling: from membrane receptors to nuclear channels. Diabetes 2004;53(1):S86-91.

61. Havel PJ. Peripheral signals conveying metabolic information to the brain: short-term and long-term regulation of food intake and energy homeostasis. Exp Biol Med 2001; 226:963-77.

62. MacDonald PE, El-Kholy W, Riedel MJ, Salapatek AM, Light PE, Wheeler MB. The multiple actions of GLP-1 on the process of glucose-stimulated insulin secretion. Diabetes 2002; 51(3):S434-42.

63. Kahn SE, McCulloch DK, Porte D. Insulin secretion in the normal and diabetic human. International Textbook of Diabetes Mellitus. (2nd ed) John Wiley & Sons, New York; 1997:337-54.

64. Liu Z, Barrett EJ. Human protein metabolism: its measurement and regulation. Am J Physiol Endocrinol Metab 2002; 283:E1105-E1112.

65. Amery CM, Nattrass M. Fatty acids and insulin secretion. Diabetes Obes Metab 2000; 2:213-21.

66. Kido Y, Nakae J, Accili D. The insulin receptor and its cellular targets. J Clin Endocrinol Metab 2001; 86:972-9.

67. Withers DJ, White M. Perspective: The insulin signaling system--a common link in the pathogenesis of type 2 diabetes. Endocrinology 2000; 141:1917-21.

68. Burks DJ, White MF. IRS proteins and beta-cell function. Diabetes 2001;50 (1)1:S140-145.

69. Hunter SJ, Garvey WT. Insulin action and insulin resistance: diseases involving defects in insulin receptors, signal transduction and the glucose transport effector system. Am J Med 1998; 105:331-45.

70. Giorgino F, Laviola L, Eriksson JW. Regional differences of insulin action in adipose tissue: insights from in vivo and in vitro studies. Acta Physiol Scand 2005;183:13- 30.

71. Karam JH. Pancreatic Hormones and Diabetes Mellitus. Basic and Clinical Endocrinology. Appleton & Lange, Stamford CT USA; 1997: 601-602.

72. Denton RM, Tavaré JM. Molecular basis of insulin action on intracellular metabolism. International Textbook of Diabetes Mellitus (2nd ed) John Wiley & Sons, New York; 1997: 469-88.

73. World Health Organization Laboratory Diagnosis and Monitoring of Diabetes Mellitus. 2002.

74. Klotzsch SG, McNamara JR. Triglyceride Measurements: A Review of Methods and Interferences. Clin Chem 1990; 36(9):1605-1613.

75. Nauck M, Marz W, Haas B, Wieland H. Homogeneous assay for direct determination of high-density lipoprotein cholesterol evaluated. Clin Chem 1996;42(3): 424-429.

76. Nauck M, Warnick GR, Rifai N. Methods for Measurement of LDL-Cholesterol: A Critical Assessment of Direct Measurement by Homogeneous Assays versus Calculation. Clin Chem 2002; 48:(2):236–254.

77. Miki Y. A homogeneous assay for the selective measurement of LDL-cholesterol in serum. Enzymatic selective protection method. Clin Lab 1999;45:398-401.

78. Third Report of the National Cholesterol Education Program (NCEP) Expert Panel on NIH Publication No. 02-5215 2002:I1-IX7.

79. Hahn RG, Ljunggren S, Larsen F , Nyström T. A simple intravenous glucose tolerance test for assessment of insulin sensitivity. Theoretical Biology and Medical Modelling 2011;8:12.

80. Rubenstein A, Clark J, Melani F, Steiner D. Secretion of proinsulin, C-peptide by pancreatic beta cells and its circulation in blood. Nature 1969; 224:697–699.

81. Steiner D, Cunningham D, Spigelman L, Aten B. Insulin biosynthesis: evidence for a precursor. Science 1967; 157:697–700.

82. Steiner D, Oyer P. Biosynthesis of insulin and a probable precursor of insulin by a human islet cell adenoma. Proc Natl Acad Sci 1967; 57:473–481.

83. Steiner D, Bell G, Tager H. Chemistry and biosynthesis of pancreatic protein hormones. J Pediatr Endocrinol 1995:1296–1328.

84. Hendriksen C, Faber OK, Drejer J, Binder C. Prevalence of residual 1-cell function in insulin treated diabetics evaluated by the plasma C-peptide response to intravenous glucagon. Diabetologia 1997; 13:615-619.

85. Hoekstra JBL, Van Rijn HJM, Erkelens DW, Thijssen JHH. C-peptide. Diabetes Care 1982; 5:438-446.

86. Scarlett JA, Mako ME, Rubenstein AH, Blix PM, Goldman J, Horwitz DL, et al. Factitious hypoglycemia: diagnosis by measurement of serum C-peptide immunoreactivity and insulin binding antibodies. N Engl J Med 1977; 297:1029-1032.

87. Rossell R , Yomis R, Casamitjana R, Segura R, Vilardell E, Rivera F. Reduced hepatic insulin extraction in obesity: relationship with plasma insulin levels. J Clin Endocrinol Metab 1983; 56:608-611.

88. Meistas MT, Margolis S, Kowarski AA. Hyperinsulinemia of obesity is due to decreased clearance of insulin. Am J Physiol 1983; 245:E155-E159.

89. Bonora E, Zavaroni I, Coscelli C, Butturini N. Decreased hepatic insulin extraction in subjects with mild glucose intolerance. Metabolism 1983; 32:438-446.

90. Savage PJ, Flock EV, Mako ME, Blix PM , Rubenstein AH, Bennett PH. C-peptide and insulin secretion in Pima Indians and Caucasians: constant fractional hepatic extraction over a wide range of insulin concentrations and in obesity. J Clin Endocrinol Metab 1979;48:594-598.

91. Sando H, Lee SV, Iwamoto T, Ikeuch M, Kosaka K. Isoproterenol stimulated C-peptide and insulin secretion in diabetic and non-obese normal subjects: decreased hepatic extraction of endogenous insulin in diabetes. J Clin Endocrinol Metab 1988; 51:1143-1149.

92. Faber OK, Christensen K, Kehlet H, Madsbad S, Binder C. Decreased insulin removal contributes to hyperinsulinemia in obesity. J Clin Endocrinol Metab 1981; 53:618-621.

93. Madsbad S, Kehlet H, Hilsted J, Troner B. Discrepancy between plasma C-peptide and insulin response to oral and intravenous glucose. Diabetes 1983; 32:436-438.

94. Lele RD. Pro-insulin, C Peptide, Glucagon, Adiponectin, TNFa, AMPK : Neglected Players in Type 2 Diabetes Mellitus. JAssoc physicians India 2010;48:1-11

95. Seaquist ER, et al. Hyperproinsulinemia is associated with increased b cell demand after hemipancreatectomy in humans J Clin Invest 1996; 97:455-60.
96. Wahren J, et al. C peptide is a bioactive peptide Diabetologia 2007; 50:503-509.

97. Wang P, Abbasi F, Carantoni M, Chen YI, Azhar S,Reaven GM. Insulin resistance does not change the ratio of proinsulin to insulin in normal volunteers. J Clin Endocrinol Metab 1997; 82:3221–3224.

98. Tura A, Ludvik B, Nolan JJ, Pacini G,Thomaseth K. Insulin and C-peptide secretion and kinetics in humans: direct and model-based measurements during OGTT. Am J Physiol Endocrinol Metab 2001; 281:E966–E974.

99. Cobelli C, Pacini G. Insulin secretion and hepatic extraction in humans by minimal modeling of C-peptide and insulin kinetics. Diabetes 1988; 37:223–232.

100. Pacini G. Mathematical models of insulin secretion in physiological and clinical investigations. Comput Methods Programs Biomed 1994;41: 269–285.

101. Thomaseth K, Kautzky-Willer A, Ludvik B, Prager R, Pacini G. Integrated mathematical model to assess β-cell activity during the oral glucose test. Am J Physiol Endocrinol Metab 1996; 270:E522–E531.

102. American Diabetes Association. Standards of medical care in diabetes-2012. Diabetes Care 2012;35:S11-S63.

103. American Diabetes Association. Diagnosis and classification of diabetes mellitus. Diabetes Care 2012;35:S64-S71.

104. Boutati E, Raptis SA, Self-Monitoring of Blood Glucose as Part of the Integral Care of Type 2 Diabetes. Diabetes Care 2009;32(2):S205–S210.

105. Diabetes Control and Complications Trial Research Group: The effect of intensive treatment of diabetes on the development and progression of long term complications in insulin-dependent diabetes mellitus. N Engl J Med 1993;329:977–986

106. American Diabetes Association: Standards of medical care in diabetes (Position Statement).Diabetes Care 2008;31(1):S12–S46

107. Alberti KG. Screening and diagnosis of prediabetes: where are we headed? Diabetes Obes Metab 2007;9(1):12– 16.

108. World Health Organization. WHO Expert Committee on Diabetes Mellitus: second report. World Health Organ Tech Reo Ser 1980; 646:1– 80.

109. American Diabetes Association. Diagnosis and classification of diabetes mellitus. Diabetes Care 2005;28(1):37– 42.

110. World Health Organization. Definition and Diagnosis of Diabetes Mellitus and Intermediate Hyperglycaemia: Report of a WHO/IDF Consultation. Geneva: World Health Organization. 2006.

111. American Diabetes Association. Diagnosis and classification of diabetes mellitus. Diabetes Care 2010;3(1):S62–69.

112. Santaguida PL, Balion C, Morrison K, et al. Diagnosis, prognosis, and treatment of impaired glucose tolerance andimpaired fasting glucose. Evidence report/technology assessment no. 128. Agency Healthcare Research and Quality Publication No 05-E026-2, 2005.

113. Shaw JE, Zimmet PZ, Alberti KG. Point: impaired fasting glucose: the case for the new American Diabetes Association criterion. Diabetes Care 2006; 29:1170-1172.

114. Forouhi NG, Balkau B, Borch-Johnsen K, et al. EDEG. The threshold for diagnosing impaired fasting glucose: a position statement by the European Diabetes Epidemiology Group. Diabetologia 2006;49:822-827.

115. Zhang X, Gregg E, Williamson D, et al. A1C level and future risk of diabetes: a systematic review. Diabetes Care 2010;33:1665-1673.

116. Heianza Y, Arase Y, Fujihara K, et al. Screening for pre-diabetes to predict future diabetes using various cut-off points for HbA1c and impaired fasting glucose: the Toranomon Hospital Health Management Center Study 4 (TOPICS 4). Diabetic Med 2012;29:e279-285.

117. Reaven GM. Banting Lecture 1988. Role of insulin resistance in human disease. Diabetes 1988;37:1595-1607.

118. Alberti KG, Eckel R, Grundy S, et al. Harmonizing the metabolic syndrome. Circulation 2009;120:1640-1645.

119.Gogia A, Agarwal PK. Metabolic syndrome. Indian journal of medical science 2006;60(2):72-81.

120. Sarafidis PA, Nilsson PM. The metabolic syndrome: a glance at its history. J Hypertens 2006;24:621–626.

121. Kylin ES. Hypertonie-Hyperglykamie-Hyperurikamie syndrome. Zentralblatt fur innere Medizin 1923;44.

122. Vague J. La differenciation sexuelle, facteur determinant des formes de l'obesite. Presse Med 1947;30:339–340.

123. Despres JP. Abdominal obesity as important component of insulin-resistance syndrome. Nutrition 1993; 9:452–459.

124. Albrink MJ, Krauss RM, Lindgrem FT, Von der GJ, Pan S, Wood PD Intercorrelations among plasma high density lipoprotein, obesity and triglycerides in a normal population. Lipids 1980;15:668– 676.

125. Ginsberg HN. Insulin resistance and cardiovascular disease. J Clin Invest 2000;106:453–458.

126. Kaplan NM. The deadly quartet. Upper -body obesity, glucose intolerance, hypertriglyceridemia, and hypertension. Arch Intern Med 1989;149:1514–1520.

127. DeFronzo RA, Ferrannini E. Insulin resistance.Amultifaceted syndrome responsible for NIDDM, obesity, hypertension, dyslipidemia and atherosclerotic cardiovascular disease. Diabetes Care 1991;14:173–194.

128. Haffner SM, Valdez RA, Hazuda HP, Mitchell BD, Morales PA, Stern MP. Prospective analysis of the insulin-resistance syndrome (syndrome X). Diabetes 1992; 41:715–722.

129. Balkau B, Charles MA. Comment on the provisional report from the WHO consultation. European group for the study of insulin resistance (EGIR). Diabet Med 1999;16:442–443.

130. Pacini G. The hyperbolic equilibrium between insulin sensitivity and secretion. Nut Metab Cardiovasc Dis 2006;16 S22–S27.

131. Bravata DM, Wells CK, Concato J, Kernan WN, Brass LM, Gulanski BI Two measures of insulin sensitivity provided similar information in a U.S. population. J Clin Epidemiol 2004;57:1214 –1217.

132. Carey DG, Jenkins AB, Campbell LV, Freund J, Chisholm DJ. Abdominal fat and insulin resistance in normal and overweight women: direct measurements reveal a strong relationship in subjects at both low and high risk of NIDDM. Diabetes 1996;45:633–638.

133. Mayer-Davis EJ, D'Agostino Jr R, Karter AJ, Haffner SM, Rewers MJ, Saad M, Bergman RN. Intensity and amount of physical activity in relation to insulin sensitivity: the Insulin Resistance Atherosclerosis Study. JAMA 1998;279:669–674.

134. Bergman RN. Lilly lecture. Toward physiological understanding of glucose tolerance. Minimal-model approach. Diabetes 1989;38:1512–1527.

135. Vessby B, Unsitupa M, Hermansen K, Riccardi G, Rivellese AA, Tapsell LC, et al. Substituting dietary saturated for monounsaturated fat impairs insulin sensitivity in healthy men and women: the KANWU Study. Diabetologia 2001; 44:312–319.

136. Bergman RN, Ader M, Huecking K, Van Citters G. Accurate assessment of β-cell function: the hyperbolic correction. Diabetes 2002;51(1):S212–S220.

137. Eckel RH, Grundy SM, Zimmet PZ. The metabolic syndrome. Lancet 2005;365:1415–1428.

138. Grundy SM. Metabolic syndrome scientific statement by the American Heart Association and the National Heart, Lung, and Blood Institute. Arterioscler Thromb Vasc Biol 2005;25:2243–2244.

139. Eckel RH, Grundy SM, Zimmet PZ. The metabolic syndrome. Lancet 2005;365:1415–1428.

140. Stefan N, Kantartzis K, Machann J, Schick F, Thamer C, Rittig K. et al. Identification and characterization of metabolically benign obesity in humans. Arch Intern Med 2008;168:1609–1616.

141. Grundy SM, Cleeman JI, Daniels SR, Donato KA, Eckel RH, Franklin BA, et al. Diagnosis and management of the metabolic syndrome. An American Heart Association/National Heart, Lung, and Blood Institute Scientific Statement. Executive summary. Cardiol Rev 2005;13:322–327.

142. Weir GC, Bonner-Weir S. Adominant role for glucose in β-cell compensation of insulin resistance. J Clin Invest 2007;117:81–83

143. Terauchi Y, Takamoto I, Kubota N, Matsui J, Suzuki R, Komeda K, et al. Glucokinase and IRS-2 are required for compensatory β- cell hyperplasia in response to high-fat diet induced insulin resistance. J Clin Invest 2007;117:246–257.

144. Goodarzi MO, Taylor KD, Guo X, Hokanson JE, Haffner SM, Cui J, Chen YD, et al. Haplotypes in the lipoprotein lipase gene influence fasting insulin and discovery of a new risk haplotype. J Clin Endocrinol Metab 2007;92:293–296.

145. Bonnycastle LL, Willer CJ, Conneely KN, Jackson AU, Burrill CP, Watanabe RM, et al. Common variants in maturity-onset diabetes of the young genes contribute to risk of type 2 diabetes in Finns. Diabetes 2006;55:2534–2540

146. Wajchenberg BL. β-Cell failure in diabetes and preservation by clinical treatment. Endocr Rev 2007;28:187–218.

147. McGarry JD. Banting lecture 2001: dysregulation of fatty acid metabolism in the etiology of type 2 diabetes. Diabetes 2002;51:7–18.

148. Randle PJ, Garland PB, Hales CN, Newsholme EA. The glucose fatty-acid cycle. Its role in insulin sensitivity and the metabolic disturbances of diabetes mellitus. Lancet 1963;1:785–789.

149. Randle PJ. Regulatory interactions between lipids and carbohydrates: the glucose fatty acid cycle after 35 years. Diabetes Metab Rev 1998;14:263–283.

150. Kahn BB, Flier JS. Obesity and insulin resistance. J Clin Invest 2000; 106:473–481.

151. Kraemer FB, Shen WJ. Hormone-sensitive lipase: control of intracellular tri-(di-)acylglycerol and cholesteryl ester hydrolysis. J Lipid Res 2002;43:1585–1594.

152. Schweiger M, Schreiber R, Haemmerle G, Lass A, Fledelius C, Jacobsen P, et al. Adipose triglyceride lipase and hormone-sensitive lipase are the major enzymes in adipose tissue triacylglycerol catabolism. J Biol Chem 2006;281:40236–40241.

153. Jocken JW, Langin D, Smit E, Saris WH, Valle C, Hul GB, Holm C, Arner P, Blaak EE. Adipose triglyceride lipase and hormone-sensitive lipase protein expression is decreased in the obese insulin-resistant state. J Clin Endocrinol Metab 2006;92:2292–2299.

154. Klannemark M, Orho M, Langin D, Laurell H, Holm C, Reynisdottir S, Arner P, et al. The putative role of the hormone-sensitive lipase gene in the pathogenesis of type II diabetes mellitus and abdominal obesity. Diabetologia 1998;41:1516–1522.

155. Kissebah AH, Vydelingum N, Murray R, Evans DJ, Hartz AJ, Kalkhoff RK, et al. Relation of body fat distribution to metabolic complications of obesity. J Clin Endocrinol Metab 1982;54:254–260.

156. Bjorntorp P. Metabolic implications of body fat distribution. Diabetes Care 14:1132–1143.

157. Pouliot MC, Despres JP, Nadeau A, Moorjani S, Prud'Homme D, Lupien PJ, et al. Visceral obesity in men. Associations with glucose tolerance, plasma insulin and lipoprotein levels. Diabetes 1992;41:826–834.

158. Arner P. Not all fat is alike. Lancet 1998;351:1301–1302.

159. Large V, Arner P. Regulation of lipolysis in humans. Pathophysiological modulation in obesity, diabetes and hyperlipidaemia. Diabetes Metab 1998;24:409–418.

160. Bergman RN, Kim SP, Hsu IR, Catalano KJ, Chiu JD, Kabir M, et al. Abdominal obesity: role in the patho- physiology of metabolic disease and cardiovascular risk. Am J Med 2007;120:S3–S8.

161. Gonzalez-Baro MR, Lewin TM, Coleman RA. Regulation of triglyceride metabolism II. Function of mitochondrial GPAT1 in the regulation of triacylglycerol biosynthesis and insulin action. Am J Physiol Gastrointest Liver Physiol 2007;292:G1195–G1199.

162. Lewis GF, Uffelman KD, Szeto LW, Steiner G. Effects of acute hyperinsulinemia on VLDL triglyceride and VLDL apo B production in normal weight and obese individuals. Diabetes 1993;42:833–842.

163. Ginsberg HN, Zhang YL, Hernandez-Ono. A Regulation of plasma triglycerides in insulin resistance and diabetes. Arch Med Res 2005;36:232–240.

164. Lewis GF, Steiner G. Acute effects of insulin in the control of VLDL production in humans.Implications for the insulin-resistant state. Diabetes Care 1996;19:390–393.

165. Sniderman AD, Faraj M. Apolipoprotein B, apolipoprotein A-I. insulin resistance and the metabolic syndrome. Curr Opin Lipidol 2007;18:633–637.

166. Murakami T, Michelagnoli S, Longhi R, Gianfranceschi G, Pazzucconi F, Calabresi L, et al. Triglycerides are major determinants of cholesterol esterification/transfer and HDL remodeling in human plasma. Arterioscler Thromb Vasc Biol 1995;15:1819–1828.

167. Lewis GF, Rader DJ. New insights into the regulation of HDL metabolism and reverse cholesterol transport. Circ Res 2005;96:1221–1232.

168. Sandhofer A, Kaser S, Ritsch A, Laimer M, Engl J, Paulweber B, et al. Cholesteryl ester transfer protein in metabolic syndrome. Obesity 2006;14:812–818.

169. Schalch DS, Kipnis DM. Abnormalities in carbohydrate tolerance associated with elevated plasma nonesterified fatty acids. J Clin Invest 1965; 44:2010–2020.
170. Boden G, Chen X, Ruiz J, White JV, Rossetti L. Mechanisms of fatty acid-induced inhibition of glucose uptake. J Clin Invest 1994;93:2438–2446.

171. Dresner A, Laurent D, Marcucci M, Griffin ME, Dufour S, Cline GW, et al. GI Effects of free fatty acids on glucose transport and IRS-1-associated phosphatidylinositol 3-kinase activity. J Clin Invest 1999;103:253–259.

172. Roden M, Krssak M, Stingl H, Gruber S, Hofer A, Furnsinn C, et al. Rapid impairment of skeletal muscle glucose transport/phosphorylation by free fatty acids in humans. Diabetes 1999;48:358–364.

173. Pan DA, Lillioja S, Kriketos AD, Milner MR, Baur LA, Bogardus C, et al. Skeletal muscle triglyceride levels are inversely related to insulin action. Diabetes 1997;46:983–988.

174. Ferrannini E, Buzzigoli G, Bonadonna R, Giorico MA, Oleggini M, Graziadei L, et al. Insulin resistance in essential hypertension. N Engl J Med 1987;317:350–357.

175. Bonora E, Capaldo B, Perin PC, Del Prato S, De Mattia G, Frittitta L. Hyperinsulinemia and insulin resistance are independently associated with plasma lipids, uric acid and blood pressure in non-diabetic subjects. The GISIR database. Nutr Metab Cardiovasc Dis 2007;18:624–631.

176. Laakso M, Sarlund H, Mykkanen L. Essential hypertension and insulin resistance in non-insulin-dependent diabetes. Eur J Clin Invest 1989; 19:518–526.

177. Steinberg HO, Brechtel G, Johnson A, Fineberg N, BaronAD. Insulin-mediated skeletal muscle vasodilation is nitric oxide dependent. A novel action of insulin to increase nitric oxide release. J Clin Invest 1994;94:1172–1179.

178. DeFronzo RA, Cooke CR, Andres R, Faloona GR, Davis PJ. The effect of insulin on renal handling of sodium, potassium, calcium, and phosphate in man. J Clin Invest 1975 55:845–855.

179. Tooke JE, HannemannMM. Adverse endothelial function and the insulin resistance syndrome. J Intern Med 2000 ;247:425–431.

180. Kuroda S, Uzu T, Fujii T, Nishimura M, Nakamura S, Inenaga T, Kimura G. Role of insulin resistance in the genesis of sodium sensitivity in essential hypertension. J Hum Hypertens 1999;13:257–262.

181. Tripathy D, Mohanty P, Dhindsa S, Syed T, Ghanim H, Aljada A, Dandona P . Elevation of free fatty acids induces inflammation and impairs vascular reactivity in healthy subjects. Diabetes 2003;52:2882–2887.

182. Grekin RJ, Vollmer AP, Sider RS. Pressor effects of portal venous oleate infusion. A proposed mechanism for obesity hypertension. Hypertension 1995;26:193–198.

183. Anderson EA, Hoffman RP, Balon TW, Sinkey CA, Mark AL. Hyperinsulinemia produces both sympathetic neural activation and vasodilation in normal humans. J Clin Invest 1991;87:2246–2252.

184. Egan BM. Insulin resistance and the sympathetic nervous system. Curr Hypertens Rep 2003;5:247–254.

185. Matthews DR, Hosker JP, Rudenski AS, Naylor BA, Treacher DF, Turner RC. Homeostasis model assessment: insulin resistance and β-cell function from fasting plasma glucose and insulin concentrations in man. Diabetologia 1985; 28:412–419.

186. Hanley AJ, Karter AJ, Festa A, D'Agostino Jr R, Wagenknecht LE, Savage P, et al. Factor analysis of metabolic syndrome using directly measured insulin sensitivity: the Insulin Resistance Atherosclerosis Study. Diabetes 2002;51:2642–2647.

187. Kim S, Soltani-Bejnood M, Quignard-Boulange A, Massiera F, Teboul M, Ailhaud G. The adipose renin-angiotensin system modulates systemic markers of insulin sensitivity and activates the intrarenal reninangiotensin system. J Biomed Biotechnol 2006:27012.

188. Vogt B, Bochud M, Burnier M. The association of aldosterone with obesity-related hypertension and the metabolic syndrome. Semin Nephrol 2007; 27:529–537.

189. Stern MP, Williams K, Gonzalez-Villalpando C, Hunt KJ, Haffner SM. Does the metabolic syndrome improve identification of individuals at risk of type 2 diabetes and/or cardiovascular disease? Diabetes Care 2004;27:2676–81.

190. Lorenzo C, Williams K, Hunt KJ, Haffner SM. The National Cholesterol Education Program–Adult Treatment Panel III, International Diabetes Federation, and World Health Organization definitions of the metabolic syndrome as predictors of incident cardiovascular disease and diabetes. Diabetes Care 2007;30:8 –13.

191. Alexander CM, Landsman PB, Grundy SM. Metabolic syndrome and hyperglycemia: congruence and divergence. Am J Cardiol 2006;98:982–5.

192. Cusi K. The role of adipose tissue and lipotoxicity in the pathogenesis of type 2 diabetes. Curr Diab Rep 2010;10:306 –315.

193. Grundy SM. Pre-Diabetes, Metabolic Syndrome, and Cardiovascular Risk JACC 2012;59(7):635–643.

194. Tenenbaum A, Fisman EZ, Michael Motro1 M, Metabolic syndrome and type 2 diabetes mellitus: focus on peroxisome proliferator activated receptors (PPAR). Cardiovascular Diabetology 2003, 2(4):1-7.

195. Balkau B, Charles MA. Comment on the provisional report from the WHO consultation. European Group for the Study of Insulin Resistance (EGIR). Diabet Med 1999;16:442–443.

196. Bloomgarden ZT. American Association of Clinical Endocrinologists (AACE) consensus conference on the insulin resistance syndrome: 25–26 August 2002, Washington, DC. Diabetes Care 2003;26:1297–1303.

197. Alberti KG, Zimmet P, Shaw J. Metabolic syndrome–a new world-wide definition. A Consensus Statement from the International Diabetes Federation. Diabet Med 2006;23:469–480.

198. DeFronzo RA, Tobin JD, Andres R. Glucose clamp technique: a method for quantifying insulin secretion and resistance. Am J Physiol 1979;237: E214- E223.

199. Singh B, Saxena A. Surrogate markers of insulin resistance: A review. World J Diabetes 2010;1(2):36-47.

200. Matsuda M, DeFronzo RA. Insulin sensitivity indices obtained from oral glucose tolerance testing: comparison with the euglycemic insulin clamp. Diabetes Care 1999;22:1462-1470.

201. Wallace TM, Levy JC, Matthews DR. Use and abuse of HOMA modeling. Diabetes Care 2004;27:1487-1495.

202. Chen H, Sullivan G, Yue LQ, Katz A, Quon MJ. QUICKI is a useful index of insulin sensitivity in subjects with hypertension. Am J Physiol Endocrinol Metab 2003;284:E804-E812.

203. Radziuk J. Insulin sensitivity and its measurement: structural commonalities among the methods. J Clin Endocrinol Metab 2000;85:4426-4433.

204. Duncan MH, Singh BM, Wise PH, Carter G, Alaghband- Zadeh J. A simple measure of insulin resistance. Lancet 1995;346:120-121.

205. Hrebícek J, Janout V, Malincíková J, Horáková D, Cízek L. Detection of insulin resistance by simple quantitative insulin sensitivity check index QUICKI for epidemiological assessment and prevention. J Clin Endocrinol Metab 2002; 87:144-147.

206. Diagnosis and classification of diabetes mellitus. Diabetes Care 2007; 30(1)1:S42-S47.

207. Dalla MC, Campioni M, Polonsky KS, Basu R, Rizza RA, Toffolo G, Cobelli C. Two-hour seven-sample oral glucose tolerance test and meal protocol: minimal model assessment of beta-cell responsivity and insulin sensitivity in nondiabetic individuals. Diabetes 2005;54:3265-3273.

208. Hsieh CH, Kuo SW, Hung YJ, Shen DC, Ho CT, Lian WC, Lee CH, Fan SC, Pei D. Metabolic characteristics in individuals with impaired glucose homeostasis. Int J Clin Pract 2005;59:639-644.

209. Consensus Development Conference on Insulin Resistance. 5th-6th November 1997. American Diabetes Association. Diabetes Care 1998;21:310-314.

210.Tripathi B, Srivastava A. Diabetes mellitus: Complications and therapeutics. Med Sci Monit 2006; 12(7): RA130-147.

211. Aiello LP, Gardner TW, King GL, et al. Diabetic retinopathy. Diabetes Care 1998; 21:143–56.

212. Chen YD, Reaven GM. Insulin resistance and atherosclerosis. Diabetes Rev 1997; 5: 331–343.

213. Ritz E, Orth SR. Nephropathy in patients with type 2 diabetes. N Engl J Med 1999; 341:1127–1133.

214. Grundy SM, Benjamin IJ, Burke GL, et al. Diabetes and cardiovascular disease: A statement for healthcare professionals from the American Heart Association. Circulation 1999;100:1134–1146.

215. Fox CS, Coady S, Sorlie PD et al. Trends in cardiovascular complications of diabetes. JAMA 2004;292:2495–2499.

216. Winer N, Sowers JR. Epidemiology of diabetes. J Clin Pharmacol 2004; 44: 397–405.

217. Duby JJ, Campbell RK, Setter SM, White J R, Rasmussen KA. Diabetic neuropathy: an intensive review. Am J Health Syst Pharm 2004;61:160–173.

218. Goldberg RB. Cardiovascular disease in patients who have diabetes. Cardiol Clin 2003;21:399–413.

219. Kikkawa R, Koya D, Haneda M. Progression of diabetic nephropathy. Am J Kidney Dis 2003;41:S19–S21.

220. Porta M, Bandello F. Diabetic retinopathy: a clinical update. Diabetologia 2002;45:1617–1634.

221. De Caterina R. Endothelial dysfunctions: common denominators in vascular disease. Curr Opin Lipidol 2000;11:9–23.

222. Stehouwer CD, Lambert J, Donker AJ, van Hinsbergh VW. Endothelial dysfunction and pathogenesis of diabetic angiopathy. Cardiovasc Res 1997;34: 55–68.

223. Cines DB, Pollak ES, Buck CA, et al. Endothelial cells in physiology and in the pathophysiology of vascular disorders. Blood 1998;91:3527–3561.

224. Quyyumi AA. Endothelial function in health and disease: new insights into the genesis of cardiovascular disease. Am J Med 1998; 105:32S–39S.

225. Kawashima S. The two faces of endothelial nitric oxide synthase in the pathophysiology of atherosclerosis. Endothelium 2004;11:99–107.

226. Khan BV, Harrison DG, Olbrych MT, Alexander RW. Medford RM. Nitric oxide regulates vascular cell adhesion molecule 1 gene expression and redox-sensitive transcriptional events in human vascular endothelial cells. Proc Natl Acad Sci 1996;93:9114–9119.

227. Janssen-Heininger YM, Poynter ME, Baeuerle PA. Recent advances towards understanding redox mechanisms in the activation of nuclear factor κB. Free Radicals Biol Med 2000; 28:1317–1327.

228. Caballero AE. Endothelial Dysfunction in Obesity and Insulin Resistance: A Road to Diabetes and Heart Disease. Obesity Research 2003;11(11):1278-1289.

229. Williams SB, Cusco JA, Roddy MA, Johnstone MT, Creager MA. Impaired nitric-oxide mediated vasodilation in patients with non-insulin dependent diabetes mellitus. J Am Coll Cardiol 1996;27:567–74.

230. Lim SC, Caballero AE, Smakowski P, LoGerfo FW, Horton ES, Veves A. Soluble intercellular adhesion molecule, vascular cell adhesion molecule, and impaired microvascular reactivity are early markers of vasculopathy in type 2 diabetic individuals without microalbuminuria. Diabetes Care 1999;22:1865–70.

231. King GL, Wakasaki H. Theoretical mechanisms by which hyperglycemia and insulin resistance could cause cardiovascular diseases in diabetes. Diabetes Care 1999;22(3):C31–37.

232. Murad F. Discovery of some of the biological effects of nitric oxide and its role in cell signaling. Angew Chem Int Ed 1999;38:1856-1868.

233. Furchgott RF: Endothelium-derived relaxing factor: discovery, early studies, and identification as nitric oxide. Angew Chem Int Ed 1999 ;38:1870-1880.

234. Ignarro LI: NO in vascular biology. Angew Chem Int Ed 1999; 38:1882-1892.

235. Hemmens B, Mayer B. Enzymology of nitric oxide synthases. Methods in molecular biology. Nitric Oxide protocols, 1st ed, Totowa, NJ: Humana Press, 1998;1–32

236. Heme-dioxygen intermediate in the first and second steps of nitric oxide synthesis as studied by stopped-flow spectroscopy under single-turnover conditions. Biochemistry 2000;39:2332-2339.

237. Pufahl RA, Nanjappan PG, Woodard RW, Marletta MA. Mechanistic probes of N-hydroxylation of L-arginine by the inducible nitricoxide synthase from murine macrophages. Biochemistry 1992;31:6822-6828.

238. Stuehr DS, Kwon NS, Nathan CF, Griffith OW, Feldman PL, Wiseman J: Nw-hydroxy-L-arginine is an intermediate in the biosynthesis of nitric oxide from L-arginine. J Biol Chem 1991;266:6259-6263.

239. Korth HG, Sustmann R, Thater C, Butler AR, Ingold KU: On the mechanism of the nitric-oxide synthase-catalyzed conversion of N-omega-hydroxy-L-arginine to citrulline and nitric-oxide. J Biol Chem 1994;269:17776-17779.

240. Hevel JM, Marletta MA. Macrophage nitric oxide synthase: tetrahydrobiopterin decreases the NADPH stoichiometry. Adv Exp Med Biol 1993; 338:285-288.

241 John T Groves, Charles C-Y Wang. Nitric oxide synthase: models and mechanisms. Current Opinion in Chemical Biology 2000;4(6):687–695.

242. Ghafourifar P, Richter C. Nitric oxides synthase activity in mitochondria. FEBS Lett 1997;418:291-296.

243. Ghafourifar P, Schenk U, Klein SD, Richter C: Mitochondrial nitricoxide synthase stimulation causes cytochrome C release from isolated mitochondria evidence for intramitochondrial peroxynitrite formation. J Biol Chem 1999; 274:31185-31188.

244..Matter H, Kotsonis P. Biology and Chemistry of the Inhibition of Nitric Oxide Synthases by Pteridine-Derivatives as Therapeutic Agents. Medicinal Research Reviews 2004; 24(5):662-684.

245.Moncadab S, Higgs A, Furchgott R. International Union of Pharmacology Nomenclature in Nitric Oxide Research. Pharmacological Reviews 1997;49(2):138-142.

246.Charles IG, Palmer RM J, Hickery MS, Bayliss MT, Chubb A P, Hall VS, Moss DW, Moncada S.: Cloning, characterization and expression of a cDNA encoding an inducible nitric oxide synthase from the human chondrocyte. Proc Natl Acad Sci 1993;90:11419–11423.

247.Marsden PA, Schappert KT, Chen HS, Flowers M, Sundell CL, Wilcox JN, Lamas S, Michel T. Molecular cloning and characterization of human endothelial nitric oxide synthase. FEBS Lett. 1992;307:287–293

248. Nakane M, Schmidt HW, Pollock JS, Forstermann U, Murad F: Cloned human brain nitric oxide synthase is highly expressed in skeletal muscle. FEBS Lett 1993;316:175–180.

249. Huang PL, Fishman MC: Genetic analysis of nitric oxide synthase isoforms: targeted mutation in mice. J Mol Med 1996;74:415–421.

250. Radomski MW, Palmer RMJ, Moncada S. An L-arginine/nitric oxide pathway present in human platelets regulates aggregation. Proc Natl Acad Sci 1990; 87:5193–5197.

251. Dinerman J, Dawson TM, Schell MJ, Snowman A, Snyder. H. Endothelial nitric oxide synthase localized to hippocampal pyramidal cells: implications for synaptic plasticity. Proc Natl Acad Sci 1994;91:4214–4218.

252.Kobzik L, Bredt DS, Lowenstein CJ, Drazen J, Gaston B, Sugarbaker D, Stamler JS: Nitric oxide synthase in human and rat lung: immunocytochemical and histochemical localization. Am J Respir Cell Mol Biol 1993;9:371–377.

253. Sessa WC, Pritchard K, Seyedi N, Wang J, Hintze TH: Chronic exercise in dogs increases coronary vascular nitric oxide production and endothelial cell nitric oxide synthase gene expression. Circ Res 1994;74:349–353.

254.Weiner CP, Lizasoain I, Baylis SA, Knowles R G, Charles IG, Moncada S. Induction of calcium-dependent nitric oxide synthases by sex hormones. Proc Natl Acad Sc 1994;91:5212–5216.

255.Baylis SA, Weiner CP, Moncada S, Charles IG: In vivo expression of inducible nitric oxide synthase in the human fetus (abstract 3). Presented at the First International Conference on Biochemistry and Molecular Biology of Nitric Oxide, Los Angeles, California, 1994:1966.

256. http://www.genecards.org accessed on 24.09.2013

257. Marsden PA, Heng HH, Scherer SW, Stewart RJ, Hall AV, Shi XM, et al. Structure and chromosomal localization of the human constitutive endothelial nitric oxide synthase gene. J Biol Chem 1993;268:17478-17488

258. Zhang Q, Church JE, Jagnandan D, Catravas JD, Sessa WC, Fulton D. Functional relevance of Golgi and plasma membrane-localized endothelial NO synthase in reconstituted endothelial cells. Arterioscler Thromb Vasc Biol 2006; 26:1015-1021.

259. Sessa WC. eNOS at a glance. J Cell Sci 2004;117:2427-2429.

260. Boo YC, Kim HJ, Song H, Fulton D, Sessa W. Coordinated regulation of endothelial nitric oxide synthase activity by phosphorylation and subcellular localization. Free Radic Biol Med 2006;41(1):144-53.

261. Fulton D, Gratton JP, McCabe TJ, Fontana J, Fujio Y, Walsh K, et al. Regulation of endothelium-derived nitric oxide production by the protein kinase Akt. Nature 1999;399:597-601.

262. Dimmeler S, Fleming I, Fisslthaler B, Hermann C, Busse R, Zeiher AM. Activation of nitric oxide synthase in endothelial cells by Akt-dependent phosphorylation. Nature 1999;399: 601-605.

263. Mount PF, Kemp BE, Power DA. Regulation of endothelial and myocardial NO synthesis by multi-site eNOS phosphorylation. J Mol Cell Cardiol 2007;42: 271-219.

264. Hingorani AD, Liang CF, Fatibene J, Lyon A, Monteith S, Parsons A, et al. A common variant of the endothelial nitric oxide synthase (Glu298-->Asp) is a major risk factor for coronary artery disease in the UK. Circulation 1999;100: 1515-1520.

265. Hibi K, Ishigami T, Tamura K, Mizushima S, Nyui N, Fujita T, et al. Endothelial nitric oxide synthase gene polymorphism and acute myocardial infarction. Hypertension 1998;32:521-526.

266. Shimasaki Y, Yasue H, Yoshimura M, Nakayama M, Kugiyama K, Ogawa H, et al. Association of the missense Glu298Asp variant of the endothelial nitric oxide synthase gene with myocardial infarction. J Am Coll Cardiol 1998; 31: 1506-1510.

267. Miyamoto Y, Saito Y, Kajiyama N, Yoshimura M, Shimasaki Y, Nakayama M, et al. Endothelial nitric oxide synthase gene is positively associated with essential hypertension. Hypertension 1998;32:3-8.

268. Lacolley P, Gautier S, Poirier O, Pannier B, Cambien F, Benetos A. Nitric oxide synthase gene polymorphisms, blood pressure and aortic stiffness in normotensive and hypertensive subjects. J Hypertens 1998; 6:31-5.

269. Karantzoulis-Fegaras F, Antoniou H, Lai SL, Kulkarni G, D'Abreo C, Wong GK, et al. Characterization of the human endothelial nitric-oxide synthase promoter. J Biol Chem 1999;274:3076-93.

270. Poirier O, Mao C, Mallet C, Nicaud V, Herrmann SM, Evans A, et al. Polymorphisms of the endothelial nitric oxide synthase gene- no consistent association with myocardial infarction in the ECTIM study. Eur J Clin Invest 1999;29:284-290.

271. Miyahara K, Kawamoto T, Sase K, Yui Y, Toda K, Yang LX, et al. Cloning and structural characterization of the human endothelial nitric-oxide-synthase gene. Eur J Biochem 1994;223:719-726.

272. Philip I, Plantefeve G, Vuillaumier-Barrot S, Vicaut E, LeMarie C, Henrion D, et al. G894T polymorphism in the endothelial nitric oxide synthase gene is associated with an enhanced vascular responsiveness to phenylephrine. Circulation 1999;99:3096-3098.

273. Hingorani AD. Polymorphisms in endothelial nitric oxide synthase and atherogenesis: John French Lecture 2000. Atherosclerosis 2001;154:521-527.

274. Yoshimura M, Yasue H, Nakayama M, Shimasaki Y, Sumida H, Sugiyama S, et al. A missense Glu298Asp variant in the endothelial nitric oxide synthase gene is associated with coronary spasm in the Japanese. Hum Genet 1998;103: 65-69.

275. Kato N, Sugiyama T, Morita H, Nabika T, Kurihara H, Yamori Y, et al. Lack of evidence for association between the endothelial nitric oxide synthase gene and hypertension. Hypertension 1999; 33: 933-6.

276. Benjafield AV, Morris BJ. Association analyses of endothelial nitric oxide synthase gene polymorphisms in essential hypertension. Am J Hypertens 2000; 13: 994-998.

277. Wang XL, Wang J. Endothelial nitric oxide synthase gene sequence variations and vascular disease. Mol Genet Metab 2000; 70: 241-251.

278. Wang XL, Sim AS, Wang MX, Murrell GA, Trudinger B, Wang J. Genotype dependent and cigarette specific effects on endothelial nitric oxide synthase gene expression and enzyme activity. FEBS Lett 2000;471:45-50.

279. Vallance P, Collier J, Moncada S. Effects of endothelium-derived nitric oxide on peripheral arteriolar tone in man. Lancet 1989; 2:997-1000.

280. Haynes WG, Noon JP, Walker BR, Webb DJ. Inhibition of nitric oxide synthesis increases blood pressure in healthy humans. J Hypertens 1993;11:1375-1380.

281. Frandsenn U, Bangsbo J, Sander M, Hoffner L, Betak A, Saltin B, et al. Exercise-induced hyperaemia and leg oxygen uptake are not altered during effective inhibition of nitric oxide synthase with N(G)-nitro-L-arginine methyl ester in humans. J Physiol 2001; 531:257-64.

282. Fabi F, Argiolas L, Chiavarelli M, Del Basso P. Nitric oxide-dependent and -independent modulation of sympathetic vasoconstriction in the human saphenous vein. Eur J Pharmacol 1996;309:41-50.

282. Haffner SM, Mykka¨nen L, Festa A, Burke JP, Stern MP. Insulin-Resistant prediabetic subjects have more atherogenic risk factors than insulin-sensitive prediabetic subjects implications for preventing coronary heart disease during the prediabetic state. Circulation 2000;101:975-980.

283. Katsuki A, Sumida Y, Gabazza Ec, Murashima S, et al. Homeostasis Model Assessment is a reliable indicator of insulin resistance during follow-up of patients with type 2 diabetes. Diabetes Care 2001;24:362–365.

284. Laaksonen DE, Lakka H, Niskanen LK, Kaplan GA, Salonen JT, Lakka TA. Metabolic syndrome and development of diabetes mellitus: Application and validation of recently suggested definitions of the metabolic syndrome in a prospective cohort study. Am J Epidemiol 2002;156(11):1-8.

285.Yokoyama H, Emoto M, Fujiwara S, Motoyama K, Morioka T, Komatsu M, et al. Quantitative Insulin Sensitivity Check Index and the Reciprocal Index of Homeostasis Model Assessment in normal range weight and moderately obese Type 2 Diabetic Patients.Diabetes Care 2003;26:2426–2432.

286. Park YW, Zhu S, Palaniappan L, Heshka S, Carnethon MR, Heymsfield SB. The metabolic syndrome: prevalence and associated risk factor findings in the US population from the Third National Health and Nutrition Examination Survey, 1988-1994, Arch Intern Med 2003;163:427-436.

287. Monti LD, Barlassina C, Citterio L, Galluccio E, Berzuini C, Setola E. et al. Endothelial Nitric Oxide Synthase Polymorphisms are associated with type 2 diabetes and the insulin resistance syndrome. Diabetes 2003;52:1270-1275.

288. Hoffmann IS, Tavares-Mordwinkin R, Castejon AM, Alfieri AB, Cubeddu. LX. Endothelial nitric oxide synthase polymorphism, nitric oxide production, salt sensitivity and cardiovascular risk factors in Hispanics. J of Hum Hypertens 2005; 19:233–240

289.Lu JX, Bao YQ, Jia WP, Xiang KS. Characteristic of hypertensive subjects with metabolic syndrome and its components in communities. Acta Academiae Medicinae Sinicae 2006;28(6):756-60.

290. Nishevitha NS, Angeline T,Jeyaraj N. Endothelial nitric oxide synthase (eNOS) Glu298àAsp polymorphism (G894T) among south Indians. Indian J Med Res 2009;129:68-71

291. Ahluwalia TS, Ahuja M, Rai TS, Kohli HS, Sud K, Bhansali A, Khullar M Endothelial nitric oxide synthase gene haplotypes and diabetic nephropathy among Asian Indians. Mol Cell Biochem 2008;14(1-2):9-17.

292. Odeberg J, Larsson CA, Råstam L, Lindblad U. The Asp298 allele of endothelial nitric oxide synthase is a risk factor for myocardial infarction among patients with type 2 diabetes mellitus. BMC Cardiovascular Disorders 2008;8:36.

293. Yang B, Xu JR, Liu XM, Yang Y, Na XF, Li M, Wang YJ. Polymorphisms of rs1799983 (G>T) and rs1800780 (A>G) of the eNOS gene associated with susceptibility to essential hypertension in the Chinese Hui ethnic population. Genet Mol Res 2013;12(3):3821-3829.

294. Ianas O. Manda D, Heltianu C, Vladoiu S, Popa P, Rosca R, Oros S, Danciulescu R. The G894T polymorphism of endothelial nitric oxide synthase gene and The endocrine-metabolic changes in metabolic syndrome: A Romanian Case control Study. Acta Endocrinologica 2009;4:447-458.

295. Chen CC, Wang WS, Chang HY, Liu JS, Chen Y. Heterogeneity of body mass index, waist circumference, and waist-to-hip ratio in predicting obesity-related metabolic disorders for Taiwanese aged 35–64 yrs. Clinical Nutrition 2009;1–6.

296.Kaur P, Radhakrishnan E, Rao SR, Sankarasubbaiyan S, Rao TV, Gupte MD.The Metabolic Syndrome and Associated Risk Factors in an Urban Industrial Male Population in South India. JAPI 2010;58:363-366.

297.Tangvarasittichai S, Poonsub P , Tangvarasittichai O, Association of serum lipoprotein ratios with insulin resistance in type 2 diabetes mellitus. Indian J Med Res 2010;131:641-664.

298.Tessari P, Cecchet D, Cosma A, Vettore M, Coracina A, Millioni R, et al. Nitric oxide synthesis is reduced in subjects with type 2 diabetes and nephropathy. Diabetes 2010;59:2152–2159

299. Hsu CC, Chang H, Huang MC, Hwang SJ, Yang YC, Tai TY, et al. Association between insulin resistance and development of microalbuminuria in type 2 diabetes. Diabetes Care 2011;34:982–987.

300. Rani PK, Raman R, Gupta A, Pal SS, Kulothungan V, Sharma T. Albuminuria and diabetic retinopathy in type 2 diabetes mellitus Sankara Nethralaya diabetic retinopathy epidemiology and molecular genetic study. Diabetology & Metabolic Syndrome 2011;3:9.

301. Sharma A, Sharma A, Agrawal A. Association of dyslipidemia with diabetic retinopathic complications in type 2 diabetes. Journal of Advance Researches in Biological Sciences 2011; (1):10-13.

302. Gosh A, Shepra ML, Bhutiya Y, Pal R, Dahal S. Serum nitric oxide status in patients with type 2 diabetes mellitus in Sikkim. Int J of applied and basic medical research 2011;1(1):1-5.

303. Mehta NN, Krishnamoorthy P, Martin SS, St Clair C, Schwartz S, Iqbal N. Usefulness of insulin resistance estimation and the metabolic syndrome in predicting coronary Atherosclerosis in type 2 diabetes mellitus. Am J Cardiol 2011; 107(3):406–411.

304. Shoukry A, Shalaby SM, Abdelazim S, Abdelazim M, Ramadan A, Ismail MI, Fouad M. Endothelial nitric oxide synthase gene polymorphisms and the risk of diabetic nephropathy in type 2 diabetes mellitus. Genet Test Mol Biomarkers 2012;16(6):574-9

305.Shah VN, Cheema BS, Kohli HS, Sharma R, Khullar M, Bhansali A. Endothelial nitric oxide synthase gene polymorphism and the Risk of diabetic neuropathy in Asian Indian patients with type 2 diabetes. J Diabetes Metab 2013;4:2.

306.Julinar, Rini EA, Bachtiar H, Oenzil F. Correlation nitric oxide level and homeostatic model assessment insulin resistance in obese adolescent. International Journal of Pediatric Endocrinology 2013;1:84.

307.Salazar MR, Carbajal HA, Espeche WG, Sisnieguez L, March CE, Balbín E. Comparison of the abilities of the plasma triglyceride/high-density lipoprotein cholesterol ratio and the metabolic syndrome to identify insulin resistance. Diab Vasc Dis Res 2013;0(0):1–7.

308.El-Mofty H, Abdel Hakim MA, Din HA, Allah OK, Mosaad PS. Retinopathy and Dyslipidemia in Type II Diabetes Mellitus in Egyptian Patients. J Clin Exp Ophthalmol 2013;4:1.

309. Kundu D, Roy A, Mandal T, Bandyopadhyay U, Ghosh E, Ray D. Relation of microalbuminuria to glycosylated hemoglobin and duration of type 2 diabetes. Nigerian Journal of Clinical Practice 2013;16(2):216-220.

310. Okita K, Iwahashi H, Kozawa J, Okauchi Y, Funahashi T, Imagawa A, Shimomura I. Homeostasis model assessment of insulin resistance for evaluating insulin sensitivity in patients with type 2 diabetes on insulin therapy. Endocrine Journal 2013; 60 (3): 283-290.

311. C.R. Kothari.Research Methodology. 2nd revised eidition.2011(reprint)

312. World Health Organization.Laboratory Diagnosis and Monitoring of Diabetes Mellitus 2002.

313. ICMR. Guidelines for Management of Type 2 Diabetes. Indian Council of Medical Research, 2005.

314. Tietz N. W., (Ed.),Textbook of Clinical Chemistry. Burtis CA and Ashwood ER, Fifth Edition, 2012.

315. Barham D, Trinder P. An improved color reagent for the determination of

blood glukose by the oxidase system. Analyst 1972;97;142-5.

316. Frier BM, Ashby JP, Nairn IM, Baris JD. Plasma insulin, C-peptide and glucagon concentration in patients with insulin-dependent diabetes treated with chlorpromide. Diab metab 1981;7(1):45-49.

317. Flier JS, Kahn CR, Roth J. Receptor, antireceptor antibodies and mechanism of insulin resistance. N Engl J Med 1979;300(8):413-419.

318. Sacks DB, carbohydrates. In: burtis CA, Ashwood ER, editors. Tietz textbook of clinical chemistry. 3rd ed. Philadelphia: W.B. Saunders Compny. 1999:790-796.

319. Jeppsson JO, Kobold U, Barr J, Finke A, et al. approved IFCC reference method for measurement of HbA1c in human blood. Clini Chem Lab Med 2002;40:78-89

320. Allain CC, Poon LS, Chan CSG, Richmond W, Fu PC. Enzymatic determination of total serum cholesterol. Clin Chem 1974;20:470-475

321. Tietz NW. Clinical guide to laboratory tests, 3rd ed. Philadelphia: WB Saunders Company. 1995:610-611.

322. Koditschek LK, Umbreit WW. Alpha-glycerophosphate oxidase in streptococcus faecium F 24. J Bacteriol 1969;98:1063-1068.

323. Trinder P. Ann Clin Biochem 1969; 6:24-27.

324. Burstein M, Scholnic H.R., Morfin R. J Lipid Res 1970

325. Friedewald WT, Levy RI,Fredrickson DS. Estimation of the concentration of low-density lipoprotein cholesterol in plasma, without use of the preparative ultracentrifuge. Clin Chem 1972;18:499–502.

326.Naiti HK, J Clin immunoassay 1986;9:155

327.Dati F, et al. Lab Med 1989;13:87

328. American Diabetes Association, "Diabetic Nephropathy", Diabetes Care 1997; 20(1):524-527

329. Boyden A, Button E, Germerog D, "Precipitin Testing With Special Reference to the Measurement of Turbidity". J Immunol 1947;57:211.

330. Hellsing K, "The Effects of Different Polymers for Enhancement of the Antigen-Antibody Reaction as Measured with Nephelometry", Protides of the Biological Fluids. 1973;23:579.

331.Miranda KM, Espey MG, Wink DA. A rapid, simple spectrophotometric method for simultaneous detection of nitrate and nitrite. Nitric Oxide 2001;5:62-71.

332. Mathew G, Glenda J, Jack L. Quantitation of nitrite and nitrate in extracellular fluids. Met Enzymol 1996;268:237-46.

333.Sambrook J, et al.Molecular Cloning: A laboratory manual, 2nd edition; Cold Spring Harbor Laboratory Pres. plainview, NY. 1989.

334.Birren B, Lai E. Pulsed Field gell electrophoresis: A practical Guide; academic press: San Diego, CA, 1993.

335. www.ncbi.nlm.nih.gov/tools/primer-blast/ accessed on 24th Sep 2013.

336. https://www.neb.com/tools-and-resources/interactive.../tm-calculator accessed on 24th Sep 2013.

337. tools.neb.com/NEBcutter2/ accessed on 24th Sep 2013.

338.Vega GL. Obesity, the metabolic syndrome and Cardiovascular disease. Am Heart J 2001;142:1108-1116.

339. American Diabetes Association. Diabetes Care 2004;27(1)

340. Petersen KF, Shulman GI. Etiology of insulin resistance. Am J Med 2006;119:S10-16.

341. Nesto RW. The relation of insulin resistance syndromes to risk of cardiovascular disease. Rev Cardiovasc Med 2003;4:S11–S18.

342. Legro RS, Castracane VD, Kauffman RP. Detecting insulin resistance in polycystic ovary syndrome: purposes and pitfalls. Obstet Gynecol Surv 2004; 59:141–154.

343.Wallace TM, Levy JC, Matthews DR. Use and abuse of HOMA modeling. Diabetes Care 2004; 27:1487–95.

344. Muniyappa R, Lee S, Chen H, Quon MJ. Current approaches for assessing insulin sensitivity and resistance in vivo: advantages, limitations, and appropriate usage. Am J Physiol Endocrinol Metab 2008;294:E15–E26.

345. Chan DC, Watts GF, Barrett PH, Mamo JCL, Redgrave TG. Markers of triglyceride-rich lipoprotein remnant metabolism in visceral obesity. Clin Chem 2002;48:278–83.

346. Marceau P, Biron S, Hould FS, Marceau S, Simard S, Thung SN, et al. Liver pathology and the metabolic syndrome X in severe obesity. J Clin Endocrinol 1999;84:1513–1517.

347. Marchesini G, Brizi M, Morselli-Labate AM, Bianchi G, Bugianesi E, McCullough AJ, et al. Association of nonalcoholic fatty liver disease with insulin resistance. Am J Med 1999;107:450–55.

348. Ryysy L, Hakkinen AM, Goto T, Vehkavaara S, Westerbacka J, et al. Hepatic fat content and insulin action on free fatty acids and glucose metabolism rather than insulin absorption are associated with insulin requirements during insulin therapy in type 2 diabetic patients. Diabetes 2000;49:749–758.

349. Phillips DI, Caddy S, Ilic V, Fielding BA, Frayn KN, Borthwick AC. et al. Intramuscular triglyceride and muscle insulin sensitivity: evidence for a relationship in nondiabetic subjects. Metabolism 1996;45:947–950.

350. Goodpaster BH, Thaete FL, Simoneau JA, Kelley DE. Subcutaneous abdominal fat and thigh muscle composition predict insulin sensitivity independently of visceral fat. Diabetes 1997;46:1579–1585.

351. Pan DA, Lillioja S, Kriketos AD, Milner MR, Baur LA, Bogardus C, et al. Skeletal muscle triglyceride levels are inversely related to insulin action. Diabetes 1997;46:983–88.

352. Krssak M, Petersen KF, Bergeron R, Price T, Laurent D, Rothman DL, et al. Intramuscular glycogen and intramyocellular lipid utilization during prolonged exercise and recovery in man: a 13C and 1H nuclear magnetic resonance spectroscopy study. J Clin Endocrinol Metab 2000;85:748–54.

353. Taskinen MR. Insulin resistance and lipoprotein metabolism. Curr Opin Lipidol 1995;6:153-60.

354. Reaven GM, Chen YD, Jeppesen J, Maheux P, Krauss RM. Insulin resistance and hyperinsulinemia in individuals with small, dense low density lipoprotein particles. J Clin Invest 1993;92:141-46.

355. Lakka HM, Laaksonen DE, Lakka TA, Niskanen LK, Kumpusalo E, Tuomilehto J, Salonen JT. The metabolic syndrome and total and cardiovascular disease mortality in middle-aged men. JAMA 2002;288:2709-16.

356. World Health Organization. Definition, diagnosis and classification of diabetes mellitus and its complications. A report of a WHO consultation. Part 1: Diagnosis and classification of diabetes mellitus. Geneva: WHO 1999:1–59.

357. Thomas EL, Fitzpatrick JA, Malik SJ, Taylor-Robinson SD, Bell JD. Whole body fat: Content and distribution. Progress in Nuclear Magnetic Resonance Spectroscopy.2013;73:56-80.

358. Nwegbu MM. Defining obesity using three anthropometric parameters, being components of two diagnostic criteria, amongst Nigerian women. Journal of Dentistry, Medicine and Medical Sciences 2012;2(1):15-18

359. American College of Sports Medicine (ACSM). Guidelines for Exercise Testing and Prescription, 7th edn. Philadelphia: Lippincott Williams & Wilkins 2006: 366.

360. Reaven G. All obese individuals are not created equal: insulin resistance is the major determinant of cardiovascular disease in overweight/obese individuals. Diabetes Vasc Dis Res 2005;2:105–12.

361. Arnl¨Ov J, Sundstr¨Om J, Ingelsson E, Lind L. Impact of BMI and the Metabolic Syndrome on the Risk of Diabetes in Middle-Aged Men. Diabetes Care 2011;34:61–65.

362. Han TS, Williams K, Sattar N, Hunt Kj, Lean MEJ, Haffner SM. Analysis of Obesity and Hyperinsulinemia in the Development of Metabolic Syndrome: San Antonio Heart Study. Obesity Research 2002;10(9):923-931.

363. Vazquez G, Duval S, Jacobs DR, Silventoinen K. Comparison of body mass index, waist circumference, and waist/hip ratio in predicting incident diabetes: a meta-analysis.Epidemiol Rev 2007;29:115-28.

364.Mathur SK, Chandra P, Mishra S, Ajmera P,Sharma P. Type-2 diabetes related intermediate phenotypic traits in north Indian diabetics. Indian Journal of Clinical Biochemistry 2007;22(2):70-73.

365.McKeigue PM, Shah B, Marmot MG. Relation of central obesity and insulin resistance with high diabetes prevalence and cardiovascular risk in South Asians. Lancet 1991;337:382–386.

366. Das P, Bhattacharjee D, Bandyopadhyay S, Bhattacharya G, Singh R. Association of obesity and leptin with Insulin Resistance in type 2 diabetes mellitus in Indian population. Indian J Physiol Pharmacol 2013;57(1):45–50.

367. Cuspidi C, Meani S, Fusi V, Severgnini B, Valerio C, Catini E, Leonetti G, Magrini F, Zanchetti A. Metabolic syndrome and target organ damage in untreated essential hypertensives. J Hypertens 2004;22:1991-1998.

368. Schillaci G, Pirro M, Vaudo G, Gemelli F, Marchesi S, Porcellati C, Mannarino E: Prognostic value of the metabolic syndrome in essential hypertension. J Am Coll Cardiol 2004; 43:1817-1822.

369. Guerrero-Romero F, Rodriguez-Moran M: Concordance between the 2005 International Diabetes Federation definition for diagnosing metabolic syndrome with the National Cholesterol Education Program Adult Treatment Panel III and the World Health Organization definitions. Diabetes Care 2005;28:2588-2589.

370. Grundy SM, Brewer HB, Cleeman JI, Smith SC Jr, Lenfant C, American Heart Association; National Heart, Lung, and Blood Institute: Definition of metabolic syndrome: Report of the National Heart, Lung and Blood Institute/American Heart Association conference on scientific issues related to definition. Circulation 2004;109:433-438

371. Ferrannini E, Natali A, Capaldo B, Lehtovirta M, Jacob S, Yki-Järvinen H: Insulin resistance, hyperinsulinemia, and blood pressure: role of age and obesity. European Group for the Study of Insulin Resistance (EGIR). Hypertension 1997;30:1144-1149.

372.Millller JH, Bogdonoff MD: Antidiuresis associated with administration of insulin. J Appl Physiol 1954;6:509-512.

373. Nizet A, Lefebvre P, Crabbe J. Control by insulin of sodium potassium and water excretion by the isolated dog kidney. Pflugers Arch 1971;323:11-20.

374. Sechi LA: Mechanisms of insulin resistance in rat models of hypertension and their relationships with salt sensitivity. J Hypertens 1999;17:1229-1237.

375. Steinberg HO, Chaker H, Leaming R, Johnson A, Brechtel G, Baron AD. Obesity/insulin resistance is associated with endothelial dysfunction: implications for the syndrome of insulin resistance. J Clin Invest 1996;97: 2601–2610.

376. Ferri C, Bellini C, Desideri G, Giuliani E, De Siati L, Cicogna S, Santucci A. Clustering of endothelial markers of vascular damage in human salt sensitive hypertension. Influence of dietary sodium load and depletion. Hypertension 1998;32:862–868.

377. Fukuda N, Satoh C, Hu WY, Nakayama M, Kishioka H, Kanmatsuse K. Endogenous angiotensin II suppresses insulin signaling in vascular smooth muscle cells from spontaneously hypertensive rats. J Hypertens 2001;19:1651–1658.

378 Isenovic E, Meng Y, Divald A, Milivojevic N, Sowers JR. Role of hosphatidylinositol 3-kinase/Akt pathway in angiotensin II and insulin like growth factor-1 modulation of nitric oxide synthase in vascular smooth muscle cells. Endocrinology 2002;19:287–291.

379. Chen CH, Lin KC, Tsai ST, Chou P. Different association of hypertension and insulin-related metabolic syndrome between men and women in 8437 nondiabetic Chinese. Am J Hypertens 2000;13(7):846-853.

380. Genuth S, Alberti KG, Bennett P, Buse J, Defronzo R, Hahn R, et al.Expert Committee on the Diagnosis and Classification of Diabetes Mellitus: follow-up report on the diagnosis of diabetes mellitus. Diabetes Care 2003; 26: 3160–3167

381. Expert Committee on the Diagnosis and Classification of Diabetes Mellitus: Report of the Expert Committee on the Diagnosis and Classification of Diabetes Mellitus. Diabetes Care 1997;20:1183–1197

382. Unwin N, Shaw J, Zimmet P, Alberti KG. Impaired glucose tolerance and impaired fasting glycaemia: the current status on definition and intervention. Diabet Med 2002;19:708-723.

383. Park SH, Yoon JS, Won KC, Lee HW, Usefulness of glycated hemoglobin as diagnostic criteria for metabolic syndrome. J Korean Med Sci 2012; 27: 1057-1061.

384.Simões PP, Chagas CB, Dias VM, Castelpoggi JP, Rodrigues de Almeida RMV, Rosângela Peluso de Campos Furtado. Metabolic Syndrome and Diabetes Type 2 in Patients Cared for at a Nutrition Out-patient Facility in the City of Rio de Janeiro. J Diabetes Metab 2013; S13:1-4.

385.Bodhe C, Jankar D, Bhutada T, Patwardhan M, Patwardhan V. HbA1c: Predictor of Dyslipidemia and Atherogenicity in Diabetes Mellitus. IJBMSP 2012; 2(1):25-27.

386.Murugan K, Shrivastava DK, Patil SKB, Lanjhiyana S, Garabadu D, Ahirwar B, et al. Biochemical Investigation of Glycosylated Haemoglobin in Diabetes. Associated Nephropathy in Chhattisgarh Population. Advances in Applied Science Research 2010;1(2):106-113.

387. Scheen AJ. Treatment of type 2 diabetes. Acta Clinica Belgica, 2003;58: 318-324.
388. DeFronzo RA. Pathogenesis of type 2 diabetes mellitus: metabolic and molecular implications for identifying diabetes genes. Diabetes 1997;5:177–269.

390. Cerasi E. Insulin deficiency and insulin resistance in the pathogenesis of NIDDM: is a divorce possible? Diabetologia 1995;38:992–997.

391.Sicree RA, Zimmet P, King HO, Coventry JO. Plasma insulin response among Nauruans.Prediction of deterioration in glucose tolerance over 6 years. Diabetes 1987;36:179–186.

392. Saad MF, Knowler WC, Pettitt DJ, Nelson RG, Mott DM, Bennett PH. Sequential changes in serum insulin concentration during development of non-insulin-dependent diabetes. Lancet 1989;1(8651):1356–1359.

393. Saad MF, Knowler WC, Pettitt DJ, Nelson RG, Mott DM, Bennett PH. The natural history of impaired glucose tolerance in the Pima Indians. N Engl J Med 1988;319:1500–1505.

394. Haffner SM, Miettinen H, Gaskill SP, Stern MP. Decreased insulin secretion and increased insulin resistance are independently related to the 7-year risk of NIDDM in Mexican-Americans. Diabetes 1995;44:1386–91.

395. Weyer C, Hanson RL, Tataranni PA, Bogardus C, Pratley RE. A high fasting plasma insulin concentration predicts type 2 diabetes independent of insulin resistance. Evidence for a pathogenic role of relative hyperinsulinemia. Diabetes 2000;49:2094–2101.

396. Weyer C, Tataranni PA, Bogardus C, Pratley RE. Insulin resistance and insulin secretory dysfunction are independent predictors of worsening of glucose tolerance during each stage of type 2 diabetes development. Diabetes Care 2000;24:89–94.

397. Pimenta W, Korytkowski M, Mitrakou A, Jenssen T, Yki-Jarvinen H, Evron W, et al. Pancreatic beta-cell dysfunction as the primary genetic lesion in NIDDM. Evidence from studies in normal glucose-tolerant individuals with a first degree NIDDM relative. JAMA 1995;273:1855–1861.

398. Eriksson J, Franssila-Kallunki A, Ekstrand A, Saloranta C, Widen E, Schalin C, et al. Early metabolic defects in persons at increased risk for non-insulin-dependent diabetes mellitus. N Engl J Med 1989;321:337–343.

399. DeFronzo RA. Pathogenesis of type 2 diabetes mellitus. Med Clin N Am 2004;88:787–835.

400. Polonsky KS. The beta-cell in diabetes : from molecular genetics to clinical research. Diabetes 1995;44:705-717.

401. UKPDS Group. UK Prospective Diabetes Study 16: overview of six years' therapy of type 2 diabetes - a progressive disease. Diabetes 1995;44:1249-1258.

402.Kahn SE. The relative contributions of insulin resistance and beta-cell dysfunction to the pathophysiology of Type 2 diabetes. Diabetologia 2003;46:3-19.

403. Scheen AJ, Lefèbvre PJ. Insulin resistance vs. insulin defi ciency: which comes first ? The old question revisited. J. Wiley & Sons, New York, 2000:101-113.

404. Scheen AJ, Lefèbvre PJ. Pathophysiology of Type 2 diabetes. Handbook of Experimental Pharmacology, Oral Antidiabetics. Berlin: Springer Verlag, 1996: 7-42.

405. Polonsky KS, Sturis J, Bell GI. Non-insulin-dependent diabetes mellitus : a genetically programmed failure of the beta cell to compensate for insulin resistance. N Engl J Med 1996;334:777-83.

406. Menuet R, Lavie C, Milani R. Importance and management of dyslipidemia in the metabolic syndrome. The American Journal of the Medical Sciences 2005; 330(6):295-302.

407. Reilly M, Wolfe M, Rhodes R, Girman C, Mehta N, Rader D. Measures of insulin resistance add incremental value to the clinical diagnosis of metabolic syndrome in association with coronary atherosclerosis. Circulation 2004; 110(7):803-809.

408. Sattar N, Williams K, Sniderman A, D'Agostino R, Haffnr S. Comparison of the associations of apolipoprotein B and non-high-denisty lipoprotein cholesterol with other cardiovascular risk factors in patients with the metabolic syndrome in the Insulin Resistance Atherosclerosis Study. Circulation 2004;110 (17):2687-2693.

409. Kwiterovich PO. Clinical relevance of the biochemical, metabolic and genetic factors that influence low-density lipoprotein heterogeneity. Am J Cardiol 2002;90:30i–47i.

410. Kwiterovich PO. Identification and treatment of heterozygous familial hypercholesterolemia in children and adolescents. Am J Cardiol. 1993; 72:30D–37D.

411. Sniderman AD, Furberg CD, Keech A, Roeters van Lennep JE, Frohlich J, Jungner I, Walldius G. Apolipoproteins versus lipids as indices of coronary risk and as targets for statin treatment. Lancet 2003;361:777–780.

412. Barter PJ, Ballantyne CM, Carmena R, Castro Cabezas M, Chapman MJ, Couture P, et al., Apo B versus cholesterol in estimating cardiovascular risk and in guiding therapy: report of the thirty-person/ten-country panel. J Intern Med 2006;259:247–258.

413. Maheria P, Parchwani D, Upadhyay A, Sharma M. Analysis of Insulin Resistance in Various Components of Metabolic Syndrome. National Journal of Medical Research 2011;1:37-41.

414. Lorenzo C, Okoloise M, Williams K, Stern MP, Haffner SM. The Metabolic Syndrome as Predictor of Type 2 Diabetes. Diabetes Care 2003;26:3153–3159.

415. Marotta T, Russo BF, Ferrara LA. Triglyceride-to-HDL-cholesterol ratio and metabolic syndrome as contributors to cardiovascular risk in overweight patients. Obesity 2010;18:1608–1613.

416. Brehm A, Pfeiler G, Pacini G, Vierhapper H, Roden M. Relationship between serum lipoprotein ratios and insulin resistance in obesity. Clinical Chemistry 2004;50(12):2316–2322.

417. Knight MG, Goedecke JH, Ricks M, Evans J, Levitt NS, Marshall K, et al. The TG/HDL ratio does not predict insulin resistance in overweight women of African descent: A study of south African, African Americans and West African Women. Ethnicity & Disease 2011; 21:490-494.

418. Sung KC,Hwang ST. Association between insulin resistance and apolipoprotein B in normoglycemic Koreans. Atherosclerosis 2005 180(1):161-169.

419.Pei W, Sun Y, Lu B, Liu Q, Zhang C, Zhang J, et al. Apolipoprotein B is associated with metabolic syndrome in Chinese families with familial combined hyperlipidemia, familial hypertriglyceridemia and familial hypercholesterolemia,International Journal of Cardiology 2007;116:194–200.

420. Taskinen MR. Diabetic dyslipidaemia: from basic research to clinical practice. Diabetologia 2003;46:733–749.

421. Chahil TJ, Ginsberg HN. Diabetic dyslipidemia. Endocrinol Metab Clin North Am 2006;35:491–510.

422. Frayn KN. Adipose tissue and the insulin resistance syndrome. Proc Nutr Soc 2001; 60:375–380.

423. Adiels M, et al. Acute suppression of VLDL1 secretion rate by insulin is associated with hepatic fat content and insulin resistance. Diabetologia 2007;50:2356–2365.

424. Mooradian AD, et al. Low serum high-density lipoprotein cholesterol in obese subjects with normal serum triglycerides: the role of insulin resistance and inflammatory cytokines. Diabetes Obes Metab 2007;9:441–443.

425. Sert M, Morgul G, Tetiker BT. Diabetic dyslipidemia is a well-known issue, but what about lipoprotein a levels in Type 2 diabetics? Int J Diabetes & Metab 2010;18:81-87.

426. Kanani FH, Alam JM. Apolipoprotein B in type 2 diabetics--a cross sectional study in a tertiary care set-up. J Pak Med Assoc 2010 ;60(8):653-6.

427. Jiang R, Schulze MB, Li T, Rifai N, Stampfer MJ, Rimm EB, Hu FB. Non-HDL Cholesterol and Apolipoprotein B Predict Cardiovascular Disease Events Among Men With Type 2 Diabetes. Diabetes Care 2004;27:1991–1997.

428.Lee IT, Wang CY, Huang CN, Fu CC, Sheu WH. High triglyceride-to-HDL cholesterol ratio associated with albuminuria in type 2 diabetic subjects. J Diabetes Complication 2013;27(3):243-247.

429. Zoppini G,Negri C, Stoico V, Casati S,Pichiri I,Bonora E, Triglyceride-high-density lipoprotein cholesterol is associated with microvascular complications in type 2 diabetes mellitus. Metabolism 2012;61(1):22-9.

430. Sasongko MB, Wong TY, Nguyen TT, Kawasaki R, Jenkins A, Shaw J, et al. Serum apolipoprotein AI and B are stronger biomarkers of diabetic retinopathy than traditional lipids. Diabetes Care 2011;34:474–479.

431. Patel ML, Sachan R, Uniyal R, Gupta KK. Clinical correlation between ApoB/Apo A ratio in type 2 Diabetic Nephropathy-A Tertiary Centre Experience. International Journal of Scientific and Research Publications 2012; 2(9):1-8.

432. Caballero AE. Endothelial dysfunction in obesity and insulin resistance: a road to diabetes and heart disease. Obes Res 2003;11:1278–1289.

433. De Caterina R. Endothelial dysfunctions:common denominators in vascular disease. Curr Opin Lipidol 2000;11:9–23.

434. Stehouwer CD, Lambert J, Donker AJ, van Hinsbergh VW. Endothelial dysfunction and pathogenesis of diabetic angiopathy. Cardiovasc Res 1997; 34:55–68.

435. Sheng CS, Hu BC, Fan WX, Zou J, Li Y, Wang JG. Microalbuminuria in relation to the metabolic syndrome and its components in a Chinese population. Diabetology & Metabolic Syndrome 2011; 3:1-6.

436. Krishnaveni A, Lakshmi ANK, Paramjyothi P. Significance of microalbuminuria in metabolic syndrome. Int J Bio Med Res 2012;3(3):2093-2098.

437. Korah TE, Abdel-Atti EA, El-Edel RH. Microalbuminuria and Insulin Resistance in Patients with Type 2 Diabetes Mellitus: Association with Cardiovascular Risk Factors in Men and Women. JMRI 2005;26(4): 316-323

438. Varghese A, Deepa R, Rema M, et al. Prevalence of microalbuminuria in type 2 diabetes mellitus at a diabetes centre in southern India. Postgrad Med J 2001;77: 399-402.

439. Manaviat MR, Afkhami M, Shoja MR. Retinopathy and microalbuminuria in type II diabetic patients. . BMC Ophthalmology 2004;4:9.

440. Serne EH, IJzerman RG, De Jongh RT, Stehouwer CD Blood pressure and insulin resistance: role for microvascular function? Cardiovasc Res 2002; 55: 418–419.

441. Serne EH, IJzerman RG, Gans RO, et al. Direct evidence for insulin-induced capillary recruitment in skin of healthy subjects during physiological hyperinsulinemia. Diabetes 2002;51:1515–1522.

442. Coggins M, Lindner J, Rattigan S, et al., Physiologic hyperinsulinemia enhances human skeletal muscle perfusion by capillary recruitment. Diabetes 2001;50:2682–2690

443. Steinberg HO, Baron AD. Vascular function, insulin resistance and fatty acids. Diabetologia 2002;45:623–634.

444.Steinberg HO, Paradisi G, Hook G, Crowder K,Cronin J, Baron AD. Free fatty acid elevation impairs insulin-mediated vasodilation and nitric oxide production. Diabetes 2000;49:1231–1238.

445. Vincent MA, Clerk LH, Lindner JR, et al. Microvascular recruitment is an early insulin effect that regulates skeletal muscle glucose uptake in vivo. Diabetes 2004; 53:1418–1423.

446. Clerk LH, Rattigan S, Clark MG. Lipid infusion impairs physiologic insulin-mediated capillary recruitment and muscle glucose uptake in vivo. Diabetes 2000;51:1138–1145.

447. Brownlee M. Biochemistry and molecular cell biology of diabetic complications. Nature 2001;414:813–820.

448. Hammes H P, DuX, Edelstein D, et al. Benfotiamine blocks three major pathways of hyperglycemic damage and prevents experimental diabetic retinopathy. Nat Med 2003;9:294–299

449. Singh R, Barden A, Mori T, Beilin L. Advanced glycation end-products: a review. Diabetologia 2001;44:129–146.

450. Ginsberg HN. Insulin resistance and cardiovascular disease. J Clin Invest 2000; 106:453–458

451. Baumgartner-Parzer SM, Wagner L, Pettermann M, Grillari J, Gessl A, Waldhausl W. High-glucose–triggered apoptosis in cultured endothelial cells. Diabetes 1995; 44:1323–1327.

452. Cagliero E, Maiello M, Boeri D, Roy S Lorenzi M. Increased expression of basement membrane components in human endothelial cells cultured in high glucose. J Clin Invest 1988;82:735–738.

453. Boeri D, Almus FE, Maiello M, Cagliero E, Rao LV, Lorenzi M. Modification of tissue-factor mRNA and protein response to thrombin and interleukin 1 by high glucose in cultured human endothelial cells. Diabetes 1989;38:212–218.

454. McGinn S, Saad S, Poronnik P, Pollock CA. High glucose-mediated effects on endothelial cell proliferation occur via p38 MAP kinase. Am J Physiol Endocrinol Metab 2003;285:E708–E717.

455. Graier WF, Grubenthal I, Dittrich P, Wascher TC, Kostner GM. Intracellular mechanism of high d-glucose-induced modulation of vascular cell proliferation. Eur J Pharmacol 1995; 294:221–229.

456. Maiello M, Boeri D, Podesta F, et al. Increased expression of tissue plasminogen activator and its inhibitor and reduced fibrinolytic potential of human endothelial cells cultured in elevated glucose. Diabetes 1992;41:1009–1015.

457. Kofler S, Nickel T, Weis M. The role of cytokines in cardiovascular diseases: focus on endothelial response to inflammation. Clin Sc 2005;108:205–213.

458. Williamson JR, Chang K, Frangos M, et al. Hyperglycemic pseudohypoxia and diabetic complications. Diabetes 1993;42:801–813.

459. Williams B, Gallacher B, Patel H, Orme C. Glucose-induced protein kinase C activation regulates vascular permeability factor mRNA expression and peptide production by human vascular smooth muscle cells in vitro. Diabetes 1997; 46:1497–1503.

460. Chen S, Apostolova MD, Cherian MG, Chakrabarti S. Interaction of endothelin-1 with vasoactive factors in mediating glucose-induced increased permeability in endothelial cells. Lab Invest 2000; 80:1311–1321.

461.Yang Z, Ming XF. Recent Advances in Understanding Endothelial Dysfunction in Atherosclerosis. Clinical Medicine & Research 2005;4(1):53-65.

462.Gosh A, Shepra ML, Bhutiya Y, Pal R, Dahal S. Serum nitric oxide status in patients with type 2 diabetes mellitus in Sikkim. Int J of applied and basic medical research 2011;1(1):1-5.

463. Simão C, Lozovoy MAB, Simão C, Dichi JB, Matsuo T, Dichi I. Nitric oxide enhancement and blood pressure decrease in patients with metabolic syndrome using soy protein or fish oil. Arq Bras Endocrinol Metab 2010;54(6):540-545.

464. Dinneen SF, Gerstein HC. The association of microalbuminuria andmortality in non-insulin- dependent diabetes mellitus. A systematic overview of the literature. Arch Intern Med 1997;157:1413–1418.

465. Parving HH. Microalbuminuria in essential hypertension and diabetes mellitus. J Hypertens Suppl 1996;14:S89–S93.

466. Deckert T, Feldt-Rasmussen B, Borch-Johnsen K, et al. Albuminuria reflects widespread vascular damage. The Steno hypothesis. Diabetologia 1989;32:219-226.

467. Borch-Johnsen K, Feldt-Rasmussen B, Strandgaard S, Schroll M, Jensen JS. Urinary albumin excretion. An independent predictor of ischemic heart disease. Arterioscler Thromb Vasc Biol 1999;19:1992–1997.

468. John L, Sundar Rao PSS, Kanagasabapathy AS. Prevalence of diabetic nephropathy in non-insulin dependent diabetics. Indian J Med Res 1991;94:24-29.

469. Samanta A, Burden AC, Jagger C. A comparison of the clinical features and vascular complications of diabetes between migrant Asians and Caucasians in Leicester. Diab Res Clin Prac 1991;14:205-214.

470. Thameem F, Puppala S , Arar NH, Stern MP, Blangero J, Duggirala R, Abboud HE. Endothelial Nitric Oxide Synthase (eNOS) gene polymorphisms and their association with type 2 diabetes related traits in Mexican Americans. Diab Vasc Dis Res 2008;5(2):109–113.

471. Suryanarayana V, Rao L, Kanakavalli M, Padmalatha V, Deenadayal M, Singh L: Recurrent early pregnancy loss and endothelial nitric oxide synthase gene polymorphisms. Arch Gynecol Obstet 2006;274(2):119-124.

472. Shin SJ, Lee HH, Cha SH, Kim JH, Shim SH, Choi DH, Kim NK. Endothelial nitric oxide synthase gene polymorphisms (-786T>C, 4a4b, 894G>T) and haplotypes in Korean patients with recurrent spontaneous abortion. European journal of Obstetrics and Gynceology and Reproductive Biology. 2010;152(1): 64-67.

473. Khurana VG, Sohni YR, Mangrum WI, McClelland RL, O'Kane DJ, Meyer FB, Meissner I: Endothelial Nitric Oxide Synthase T-786C single nucleotide polymorphism: A putative genetic marker differentiating small versus large ruptured intracranial aneurysms. stroke 2003;34(11):2555-2559.

474. Senthil D, Raveendran M, Shen YH, Utama B, Dudley D, Wang J, Wang XL. Genotype dependent expression of endothelial nitric oxide synthase (eNOS) and its regulatory proteins in cultured endothelial cells. DNA Cell Biology 2005; 24(4):218-224.

475. Hingorani AD: Polymorphisms in endothelial nitric oxide synthase and atherogenesis: John French Lecture 2000. Atherosclerosis 2000;154(3):521-527.

476. Hocher B, Chen YP, Hugle S, Repey J, Krause K, Slowinski T, et al. Impact of maternal endothelial nitric oxide synthase gene polymorphisms on blood pressure, protein excretion and fetal outcome in pregnancy. J Hum Hypertens 2008;22(9):641–647.

477. Hyndman ME, Parsons HG, Verma S, Bridge PJ, Edworthy S, Jones C, Lonn E, Charbonneau F, Anderson TJ. The T-786→C Mutation in Endothelial nitric oxide synthase is associated with hypertension. Hypertension 2002;39(4): 919-922.

478. Tempfer C, Unfried G, Zeillinger R, Hefler L, Nagele F, Huber JC: Endothelial nitric oxide synthase gene polymorphism in women with idiopathic recurrent miscarriage. Human Reproduction 2001;16(8):1644-1647.

479. Karvela M, Papadopoulou S, Tsaliki E, Konstantakou E, Hatzaki A, Florentin-Arar L, Lamnissou K: Endothelial nitric oxide synthase gene polymorphisms in recurrent spontaneous abortions. Archives of Gynecology and Obstetrics 2008;278(4): 349-352.

480. Yamada H, Sata F, Saijo Y, Kishi R, Hisanori Minakami H.Genetic Factors in Fetal Growth Restriction and Miscarriage. Seminars in Thrombosis and Hemostasis 2005;31(3):334-345.

481. Makino A, Nakanishi T, Sugiura-Ogasawara M, Ozaki Y, Suzumori N, Suzumori K. No association of C677T methylenetetrahydrofolate reductase and an endothelial nitric oxide synthase polymorphism with recurrent pregnancy loss. American Journal of Reproductive Immunology 2004;52(1):60-66.

482. Shankarishan P, Borah PK, Ahmed G, Mahanta J. Prevalence of endothelial nitric oxide synthase (eNOS) gene exon 7 Glu298Asp variant in North Eastern India. Indian J Med Res 2011 133(5):487–491.

483.Sivastava K,Biswa UK, Narang R, Varghese JJ,Das N. Prevalence of eNOS Glu298Asp polymorphism in healthy volunteers from a region of Northern India. Community genet 2005;8(3):180-3.

484.Tripathi R, Arya V,Singh PK, Agarwal S. Sequence variation in eNOS (Glu298Asp) gene and its impact on coronary artery disease: North Indian Study Annals of Biological Research 2010;1(4):152-157.

485. Liu S, Yan H, Hou W, Xu P, Tian J, Tian L, Zhu B, Ma J, Lu S. Relationships between endothelial nitric oxide synthase gene polymorphisms and osteoporosis in postmenopausal women. J Zhejiang Univ Sci B 2009;10(8):609-618

486. Dimmeler S, Fleming I, Fisslthaler B, Hermann C, Busse R, Zeiher AM. Activation of nitric oxide synthase in endothelial cells by Akt-dependent phosphorylation. Nature 1999;399: 601–605

487. Verrotti A, Greco R, Basciani F, Morgese G,Chiarelli F. von Willebrand factor and its propeptide in children with diabetes. Relation between endothelial dysfunction and microalbuminuria. Pediatr Res 2003;53:382–386